PROGRESS IN HEAT TRANSFER

ISSLEDOVANIE TEPLOOBMENA V TEPLOENERGETICHESKIKH USTANOVKAKH I V USTANOVKAKH
DLYA POLUCHENIYA POLUPROVODNIKOVYKH MATERIALOV

ИССЛЕДОВАНИЕ ТЕПЛООБМЕНА В ТЕПЛОЭНЕРГЕТИЧЕСКИХ
УСТАНОВКАХ И В УСТАНОВКАХ ДЛЯ ПОЛУЧЕНИЯ
ПОЛУПРОВОДНИКОВЫХ МАТЕРИАЛОВ

PROGRESS IN
HEAT TRANSFER

Part I: Laminar Boundary Layer Flow in Transparent and Gray Media

Part II: Equipment for the Preparation of Semiconductor Materials

Edited by
P. K. Konakov

Translated from Russian by
James S. Wood

Springer Science+Business Media, LLC 1966

The Russian text was originally published as Volume 189 of
the Transactions (Trudy) of the Moscow Institute of Railroad
Engineers (MIIT) by Vysshaya Shkola Press in Moscow in 1965.

Library of Congress Catalog Card Number 65-26629

© 1966 Springer Science+Business Media New York
Originally published by Consultants Bureau in 1966.
Softcover reprint of the hardcover 1st edition 1966
A Division of Plenum Publishing Corporation
227 West 17 Street, New York, N.Y. 10011

ISBN 978-1-4899-4757-4   ISBN 978-1-4899-4755-0 (eBook)
DOI 10.1007/978-1-4899-4755-0

PREFACE TO THE AMERICAN EDITION

The present collection contains a number of articles written by members of the Department of Thermal Power Plants of the Moscow Institute of Railroad Engineers. The collection is divided into two parts. The first comprises articles relating to the theory of laminar boundary layer flow in transparent and gray media. The articles in the second part are devoted to the investigation of heat and mass transfer in equipment for the preparation of semiconductor materials.

In boundary layer theory it is assumed in analyzing stationary laminar motion of a viscous fluid about a thin plate that a thin boundary layer of variable thickness is formed on the plate, wherein the longitudinal component of the velocity varies from zero at the plate to the value of the fluid velocity in the free unperturbed flow, yet no mention is ever made in the this connection of the transverse velocity component. However, the transverse component exerts a considerable influence on the kinetics of motion about the plate due to the existence of a boundary layer associated with the transverse velocity component. In this layer the transverse component varies from zero at the plate to zero in the unperturbed flow.

Consequently, two boundary layers are formed about the plate, the first, of thickness δ, comprising the longitudinal component boundary layer, the second, of thickness Δ, comprising the transverse component boundary layer.

It is known that the boundary conditions of the Blasius problem include hydrodynamical relations at infinity. Such relations contradict the nature of the boundary layer equations, which determine the motion of the fluid only in a region of extremely small dimensions near the wall. The notion of a transverse velocity boundary layer removes this contradiction, eliminating the hydrodynamical relations at infinity from the problem in question.

The notion of a transverse velocity boundary layer also enables one to solve a number of problems relating to stationary laminar flow of a viscous fluid in channels of finite volume.

The present collection includes articles in which the kinetic processes of a viscous fluid moving past a thin plate and in a narrow slotted channel are analyzed.

In the near future articles will be published with solutions to the problems of stationary laminar motion of a viscous fluid in the initial section of a circular tube and an annular channel.

With regard to the articles in the second part of the collection, they contain only the first results of a theoretical and experimental investigation of heat and mass transfer in equipment for the preparation of semi-conductor materials. These investigations will eventually lead to results that could be of enormous practical significance.

We hope that our work will prove useful to the American reader, and we await with utmost interest the responses to the publication of our collection.

Prof. P. K. Konakov

Moscow

May 1966

PREFACE

This collection was prepared by members of the Thermal Power Equipment Department of the Moscow Institute of Railroad Engineers. The articles by P. K. Konakov give solutions to the problems of heat transfer and viscous (fluid) friction in the isothermal and nonisothermal motions of a viscous fluid with a plane laminar boundary layer.

The articles by P. K. Konakov, V. T. Kumskov, Yu. P. Sidorov, and V. S. Sidorov treat the problems of composite heat transfer and viscous friction in a moving gray medium with large and small optical densities. Analytical relations are derived for composite heat transfer in media with large and small optical densities.

The article by V. T. Kusmkov and Yu. P. Sidorov discusses methods for evaluating the absorption coefficient.

The methods of separating composite heat transfer into radiative and convective components are investigated in the articles by L. A. Goryainov. Modifications introduced recently in the separation of heat transfer in furnaces by the standardized method are analyzed.

In the article by V. S. Sidorov, an analytical solution of the problem of convective heat transfer is given on the basis of integral relations for the initial part of a tube under turbulent flow conditions, the solution agreeing satisfactorily with the experimental data.

In the article by K. F. Aksenov, V. V. Kudryavtsev, Yu. A. Lomov, A. I. Pokalyuk, and Yu. N. Khvoshchevskii the results are given of an experimental investigation of convective heat transfer and viscous friction in the motion of air in smooth pipes with Re numbers up to $1.5 \cdot 10^6$ effected at high pressures.

The article by G. E. Berevochkin and V. A. Smirnov presents data from an experimental investigation of the temperature field of germanium single crystals prepared by the Czochralski method.

The article by L. A. Zaruvinskaya and V. A. Smirnov pertains to an investigation of the temperature field in germanium single crystals with various techniques for shielding of the ingot as it is grown from a melt.

In the article by V. F. Brekhovskikh results are given from a determination of the emissive power of germanium and silicon at temperatures of 700 to 1200°K.

This collection was prepared for publication by V. T. Kumskov, V. A. Smirnov, and G. E. Verevochkin of the Thermal Power Equipment Department.

All correspondence and comments concerning the articles published in this collection should be sent to the Moscow Institute of Railroad Engineers (MIIT), ul. Obraztsova, 15, Moscow, A-55, USSR.

Prof. A. I. Ioannisyan
Dean of the Science Institute

vii

PUBLISHER'S NOTE

The following Soviet journals cited in this book are available in cover-to-cover translation:

Russian title	English title	Publisher
Kristallografiya	Soviet Physics — Crystallography	American Institute of Physics
Zhurnal Tekh- nicheskoi Fiziki	Soviet Physics — Technical Physics	American Institute of Physics

The following books cited in this volume are also available in English translation:

Growth of Crystals, Volume 2 (Izd. Akad. Nauk SSSR, 1959); English translation: New York, Consultants Bureau, 1960.

Growth of Crystals, Volume 3 (Izd. Akad. Nauk SSSR, 1960); English translation: New York, Consultants Bureau, 1962.

CONTENTS

PART I

Laminar Boundary Layer Flow in Transparent
and Gray Media

INVESTIGATIONS OF COMPOSITE HEAT TRANSFER

V. T. Kumskov

The combustion chamber operation of steam boilers, gas turbine equipment, and other specialized heat transfer apparatus is accompanied by physicochemical processes involved in combustion of the fuel and composite heat transfer.

There has been a recent surge of interest in the study of the laws of composite heat transfer, not only in the Soviet Union but in other countries as well.

In the USA this problem has been treated primarily in application to the engineering calculation of combustion chambers for power plants and of certain conditions of rocket and aircraft motion in the dense strata of the atmosphere.

The present article gives a brief review of research on composite heat transfer and indicates the basic trends of this research.

The earliest Soviet works were concerned with the theoretical and experimental investigation of combustion chambers. In 1905 Grinevetskii [1] proposed a graph-analytical method for the thermal calculation of a boiler. In 1919 there appeared the empirical method of Kirsh [2].

In 1930 Ramzin developed a method for calculating the heat transfer in furnaces [3], a method that was later developed in the work of Timofeev, Kavaderov, Pukhov, and others [4–6].

Gurvich proposed a method for calculating the heat transfer in furnaces on the basis of similarity theory [7]. After many refinements of the various coefficients, this method was eventually adopted as the method of thermal calculation norms for boiler furnaces [8].

The application of the Stefan–Boltzmann law required the introduction of certain arbitrary concepts: effective temperature of the medium and the normalized emissivity of the furnace volume. The originators of the method keep introducing "refinements," consisting in the application of additional and scarcely justified coefficients [9].

Polyak and Shorin [10, 11] analyzed special problems in radiative heat transfer in an absorbent medium. They presented a method of generalized averaging of the temperature field in furnaces [12]. However, the lack of specific data for evaluation of the coefficients, taking into account the combustion and heat transfer conditions as well as the temperature conditions in the furnace, have prevented this method from being used for practical purposes. It is shown in [13] that the heat transfer in gray bodies is determined mainly by hydrodynamic factors.

The problems of radiative heat transfer in an absorbing and scattering medium, as well as a method for the zonal calculation of radiative heat transfer in a combustion chamber, are discussed in papers by Surinov [14].

In 1950 Konakov, making use of the approximate self-similarity of the temperature field in a furnace, [15] and, subsequently, of the hypothesis of local radiative equilibrium near the wall [16], proposed a method for the thermal calculation of boiler furnaces.

Later on, a number of papers [17–22] were published which described certain investigations at the Moscow Institute of Railroad Engineers (MIIT) which were concerned with the experimental corroboration of this method.

Investigations of heat transfer in the combustion of gas and liquid fuels, as well as the effects of emissivity in the special blackening of the furnace medium, confirmed the approximate self-similarity of the temperature field in a furnace, exhibiting the qualitative dependence of composite heat transfer on the emissivity of the furnace medium. With an increase in the emissivity of the furnace medium (blackening of the medium with chrome−magnesite dust), the loss of thermal energy from the gases to the heat-sensitive surfaces was diminished.

To a first approximation, this can be attributed to the fact that when the emissivity of the furnace medium is increased, its optical density is increased, which is the reason for the reduction in photon heat transfer, i.e., increase in the fraction of absorbed radiation energy. A certain increase in the convective (molar) transfer of heat energy never completely cancels out the reduction in radiative transfer, so that the total (composite) heat transfer is diminished. The same result follows from the data of other researchers in the alternating or combined combustion of gas and liquid fuels in ovens and boiler furnaces. Consequently, the increase in emissivity of the furnace medium due to dust injection or other means of blackening the furnace medium (for example, blackening of a gaseous flame with mazut or fuel oil) does not intensify heat transfer in the furnace chamber; it diminishes it instead.

Observations have also shown that altering the emissivity of the medium largely determines the interrelation between the convective and radiative transfer of thermal energy in the furnace medium, which behaves like a gray body.

It is known that the state of a moving gray medium is characterized by the molecular (T) and radiative (T_r) temperature fields [23], which are related by the expression

$$T_r^4 - T^4 = \frac{1}{4\kappa^2} \Delta^2 T_r^4 . \tag{1}$$

where k is the absorption coefficient of the medium.

Since the molecular temperature field governs convective transfer and the radiative temperature field governs radiative transfer, it follows from Eq. (1) that with a change in the optical density (absorption coefficient) k of the medium the transfer of thermal energy is redistributed between the convective and radiative components.

Analytical solutions to special problems of radiative heat transfer in gray media have only recently begun to be worked out in the USSR and abroad [24]. One of the earliest papers in this area is an article by Smirnov [25] devoted to an investigation of heat transfer in gases by simultaneous radiation and convection in a cylindrical channel.

Later analytical investigations in composite heat transfer were conducted on the basis of a solution of the generalized boundary layer balance equation of a moving omitting medium [23, 26]; it differs from the familiar solution [27] in that the boundary layer balance equation in this case is augmented by a radiation term, taking into account the radiative heat transfer of the medium with the heat-sensitive surface. The experimental investigation of composite heat transfer in a cylindrical channel has shown that composite heat transfer depends primarily on the absorption coefficient and hydrodynamic state of the moving gray medium. The relationship between distinct types of heat transfer in a gray medium is given in [28]. There also is given an analysis of the relations obtained for special problems in radiative heat transfer.

The present collection contains a series of papers in which solutions to the laminar boundary layer equations are given for different optical densities of the medium.

The first papers by non-Soviet authors go back to a calculation of the heat transfer in boiler combustion chambers [29−34] and investigation of the emission characteristics of the combustion products [35−38].

Experimental papers have recently appeared in connection with the study of the emission characteristics of media as the pressure is varied up to 10 bar and the temperature up to 1500°K [39, 40].

In the USA many papers have been devoted to the investigation of radiative heat transfer in absorbing and emitting media. In some investigations an attempt is made to explain the mechanism of radiation energy transfer in the presence of heat conduction [41–43]. The characteristics of the temperature field with the combined effects of heat conduction and radiation are investigated.

In [44] the interaction of radiation energy and heat conduction, in a gray medium is analyzed.

In [45] a detailed investigation is conducted on the exchange of heat between a real medium (CO_2–H_2–N_2) and parallel heat-sensitive plates. The temperature of the medium was varied from 1473 to 3073°K, the Re number from 5000 to 100,000. The convective transfer was ascertained in terms of the dependences derived on the basis of boundary layer theory as a function of Re and Pr. The optical density (radiation length) RL of the medium was varied from 0.037 to 0.6 mbar.

The application of boundary layer theory for generalization of the experimental data turned out to be very rewarding, in that it became possible, by using the equivalent thermal conductivity ($\lambda_{eq} = \lambda_{mol} + \lambda_{rad}$), to obtain the following general relation for the composite heat transfer:

$$Nu = c\,Re^{0.8}Pr^{\frac{1}{3}}, \tag{2}$$

where $Nu = \alpha \cdot 2l/\lambda_{eq}$ is the Nusselt number, α is the heat transfer coefficient, l is a reference length (spacing between the walls of the channel), Pr is the Prandtl number, and λ_{eq} is the thermal conductivity coefficient, equal to the sum of the molecular and radiative thermal conductivities.

This method of generalizing the experimental data on composite heat transfer was applied in [13, 24].

In [46] the radiative transfer of heat in a spherical shell containing an absorbing and emitting gas is investigated. In this paper the gas is assumed to be a gray isotropic medium with a constant temperature. It is assumed that the absorption coefficient does not depend on the temperature.

The work reported in [47] was devoted to the simultaneous transfer of heat by heat conduction and radiation in an absorbing medium.

An analysis of the solution of a nonlinear integral equation derived by the method of successive approximations is conducted on the basis of a dimensionless parameter governing the ratio of the heat transported by heat conduction to the transfer of heat by radiation. The problem is solved by invoking an expression for the radiant heat flux vector according to Rosseland. A comparison is made between the resultant equation for the radiation flux and the formula of Shorin [11], and good agreement is obtained for gray media with a high optical density.

In [48] the effect of surface emissivity on the transfer of thermal energy is analyzed in a medium that absorbs and emits thermal radiation.

Viskanta and Grosh [49] analyze the heat transfer between an absorbing medium and a conical (wedge) heat-sensitive surface. The solution is obtained on the basis of the boundary layer equations. The authors utilize a system of boundary layer equations consisting of the equations of mass and energy conservation and the boundary layer momentum equation. The equation for the conservation of energy is simplified by means of the Rosseland radiant heat flux vector [50].

In [51] a method is examined for calculating the heat transfer by radiative emission between a gas and a gray wall. The author of the article makes use of the data of Hottel [35] to outline a well-known calculation procedure [27].

In [52] Hottel considers the fundamental problems of radiative transport. In his review he discusses [28] in detail, adding his own comments, which are of conceptual significance. We will consider these comments further.

Hottel, in delineating the diffusion concepts of radiative transport, feels that the expression given in [28] for the radiation flux vector is incorrect. The proper expression for the radiation flux vector is considered to be the expression best known in the literature as the Rosseland approximation [50].

It is known that the diffusion concepts of radiative transport permit the radiation flux vector to be represented in terms of the volume density gradient of the photon medium. It is assumed that the radiation intensity in a medium is uniform in all directions. With these assumptions the expression for the Rosseland radiation flux vector has the form

$$\vec{q}_r = -\frac{4}{3\kappa}\,\mathrm{grad}\,E, \tag{3}$$

where $E = \sigma_0 T^4$.

Hottel compares Eq. (3) with another expression for the radiation flux vector used in [28],

$$\vec{q}_r = D\,\mathrm{grad}\,\Phi, \tag{4}$$

where $D = 1/4cl_s$ is the diffusion coefficient of the photon medium, $\Phi = 4E/c = 4\sigma_0\gamma_r^4/c$ is the radiation flux density.

Comparison of Eqs. (3) and (4) reveals that they actually differ from one another only in their coefficients. In Eq. (3) the coefficient is equal to $4/3k$, and in Eq. (4) it is equal to $1/k$ if we assume that $k = 1/l_s$.

Hottel also notes that in [53] this coefficient is given the expression $4/m^2k$, where m depends on the volume distribution of the intensity. Shorin [11] showed that the variation of the spatial distribution of emissive radiation determines the various expressions for the intensity coefficient near a heat-sensitive wall; for a two-dimensional intensity distribution the coefficient is equal to $1/k$.

By analogy with the molecular kinetic theory of gases, the photon diffusion coefficient for three-dimensional flow can be written as

$$D = \frac{1}{3}\,cl_r,$$

where $l_r = 1/k$ is the mean free path of the photon and c is the velocity of light.

With a uniform three-dimensional emissive intensity distribution in all directions—which is typical of the nucleus of an absorbing medium (heart of the furnace)—the expression for the radiation flux vector has the form

$$\vec{q}_r = -\frac{1}{3}\frac{c}{\kappa}\,\mathrm{grad}\,\varepsilon_r. \tag{5}$$

Since the radiant energy density is equal to

$$\varepsilon_r = \frac{4\sigma_0 T^4}{c},$$

Eq. (5) can be written in the form of an expression analogous to the Rosseland equation:

$$\vec{q}_r = -\frac{4}{3\kappa}\,\mathrm{grad}\,E, \tag{6}$$

where $E = \sigma_0 T^4$.

For a two-dimensional emissive intensity distribution (for the region near the wall) the expression for the radiation flux vector has another form [11]:

$$\vec{q}_r = -\frac{1}{4}\frac{c}{\kappa}\,\mathrm{grad}\,\varepsilon_r. \tag{7}$$

Applying $\varepsilon_r = 4\sigma_0 T_r^4/c$, we reduce Eq. (7) to the form used in [28]:

$$\vec{q}_r = -\frac{1}{k}\,\text{grad}\ \varepsilon, \qquad (8)$$

where $E = \sigma_0 T_r^4$ and T_r^4 is the radiant temperature.

It may be concluded from the foregoing that for a wall region there is more justification for using the radiation flux vector in the form of Eq. (7). Hottel discounts the radiant temperature, first introduced by Galitsyn [54]. For an analysis of the composite transport of thermal energy it is necessary to introduce two material media: a molecular medium and a photon medium, the state of which is defined by the molecular and radiant temperatures, respectively. The ratio between these temperatures determines the deviation of the photon medium from the state of thermodynamic equilibrium. It is known that for the state of radiative equilibrium the molecular temperature is used in the analytical relations.

Hottel feels that in the analysis of radiative heat transfer problems the integral equations are the more preferable, since their solution yields more accurate final expressions than solution of the differential equations.

Both the earlier works of Polyak and his later work [55] reveal that the use of integral and differential equations gives almost identical results. The essential problem in this case is the determination of which expression to use for the radiant energy transfer vector.

In the analysis of radiative transport problems in an absorbing medium a very important quantity is the absorption coefficient of the medium, k. In most studies k is understood to mean a certain coefficient of proportionality that results from Burger's law, which takes into account the effect of directional attenuation of radiant energy when the latter is transmitted through an absorbing medium. This concept of the absorption coefficient is not always discerned, and this is sometimes the reason for contradictions between the end results of radiative transfer in absorbing media.

Our brief survey of papers on composite heat transfer leads to certain conclusions as to the trends in research on composite heat transfer problems:

1. The diffusion concepts of radiative transport in absorbing media are fruitful and are used by the majority of researchers.

2. The expressions for the radiant energy transport vector must be used with allowance for the characteristics of the volume distribution of radiation intensity.

3. The application of integral and differential equations for the problems of radiative heat transfer yield essentially the same results. Reliable results are also obtained with the application of boundary layer equations for radiative heat transfer problems in absorbing media.

4. Little experimental research has been done on the mechanism of simultaneous radiative and convective transfer of thermal energy for different optical density characteristics of the medium. There is a need for projects to find out the analytical values of the absorption coefficient.

The arguments could provide the basis for the derivation of analytical relations for composite heat transfer in the combustion chambers of furnaces, boilers, and in heat exchangers, wherein the transport of thermal energy is accomplished by the simultaneous action of heat conduction, convection, and radiation.

Literature Cited

1. V. I. Grinevetskii, Graphical Calculation of a Boiler, Byull. Politekhn. Obshch., No. 4 (1905).
2. K. V. Kirsh, Wood as a Fuel (Izd. Politekhn. Obshch., 1919).
3. L. K. Ramzin, Radiative Emission in Boiler Equipment, Izv. VTI, No. 4 (57) (1930).
4. V. N. Timofeev, Radiative Heat Transfer in the Combustion Chamber, Izv. VTI, No. 9 (87) (1934). Heat Transfer in the Combustion Chamber, Izv. VTI, No. 2 (1954).

5.   A. V. Kavaderov, Thermal Operation of Metallurgical Flame Furnaces (Metallurgizdat, 1956).

6.   V. I. Pukhov, Theoretical Procedure for Calculation of Heat Transfer in Furnaces from the Effective Radiation Temperature, Collected Scientific Papers of the Ivanovsk Power Institute, No. 6 (1955).

7.   A. M. Gurvich, General Empirical Method for Calculating the Direct Emission of Furnaces, Collected Materials of the Council for Science and Technology (NTS), No. 2 (1933). Application of Similarity Theory to Radiative Heat Transfer in Furnaces, Dokl. Akad. Nauk SSSR (new series), Vol. 27, No. 8 (1940). Heat Exchange in Boiler Furnaces (Gosénergoizdat, 1950).

8.   Thermal Calculation of Boiler Systems; Standard Method (Gosénergoizdat, 1957).

9.   A. M. Gurvich and V. V. Mitor, Calculation of Heat Transfer in Mazut-Gas and Coal Dust Furnaces, Énergomashinostroenie, No. 2 (1963).

10. G. L. Polyak, Radiative Heat Transfer in the Presence of a Radiation Absorbing and Scattering Medium, Dokl. Akad. Nauk SSSR, Vol. 27, No. 1 (1940).

11. S. N. Shorin, Heat Transfer in an Absorbing Medium (Doctoral Thesis, 1950), Izv. Akad. Nauk SSSR, No. 3 (1951).

12. G. L. Polyak and S. N. Shorin, Theory of Heat Transfer in Furnaces, Izv. Akad. Nauk SSSR, Otd. Tekhn. Nauk, No. 12 (1949).

13. V. N. Adrianov and S. N. Shorin, Heat Transfer in a Flow of Emitting Combustion Products in a Channel, Teploénergetika, No. 3 (1957).

14. Yu. A. Surinov, Radiative Transfer in the Presence of an Absorbing and Scattering Medium, Izv. Akad. Nauk SSSR, Otd. Tekhn. Nauk, Nos. 9-10 (1952). Method of Zonal Calculation of Radiative Heat Transfer in a Combustion Chamber, Izv. Akad. Nauk SSSR, Otd. Tekhn. Nauk, No. 7 (1953).

15. P. K. Konakov, Thermal Emission in a Boiler Furnace, Izv. Akad. Nauk SSSR, Otd. Tekhn. Nauk, No. 6 (1950). Self-Similarity of the Temperature Field in Boiler Furnaces, Coll.: Similarity Theory and Simulation (Izd. Akad. Nauk SSSR, 1951).

16. P. K. Konakov, S. S. Filimonov, and B. A. Khrustalev, Calculation of Heat Transfer in Boiler Furnaces, Teploénergetika, No. 8 (1957). Calculation of Radiative Heat Transfer in Cooled Combustion Chambers, Zh. Tekhn. Fiz., 27, No. 5 (1957). Heat Transfer in Steam Boiler Combustion Chambers (Rechizdat, 1960).

17. V. K. Orlov, Effect of the Lining of a Combustion Chamber on the Heat Emission in Combustion of a Gas, Proceedings of the Fourth Conference of Junior Scientists (Izd. ÉNIN Akad. Nauk SSSR, 1958).

18. L. A. Goryainov, Investigation of Composite Heat Transfer in a Cooled Channel, Tr. LIIZhTa (Transactions of the Leningrad Railroad Engineers Institute), No. 160 (Transzheldorizdat, 1958).

19. V. T. Kumskov and V. S. Sidorov, Calculation of Heat Transfer in Boiler Furnaces, Tr. MIITa (Transactions of the Moscow Railroad Engineers Institute), No. 125 (Transzheldorizdat, 1960).

20. S. M. Pokrovskii, Composite Heat Transfer in Liquid Fuel Combustion Chambers, Tr. MIITa, No. 125 (Transzheldorizdat, 1960).

21. V. T. Kumskov, Investigation of Composite Heat Transfer in Combustion Chambers, Coll.: Similarity Theory and Its Application in Heat Engineering, Proceedings of the First Inter-Institutional Conference, No. 139 (Transzheldorizdat, 1961).

22. V. I. Lebedev, Effect of the Emissivity of the Furnace Medium on Heat Transfer in Combustion Chambers, Coll.: Similarity Theory and Its Application in Heat Engineering, Proceedings of the First Inter-Institutional Conference, No. 139 (Transzheldorizdat, 1961).

23. P. K. Konakov, Certain Laws of Composite Heat Transfer, Collected Papers of MIIT, No. 125 (Transzheldorizdat, 1960). Coll.: Similarity Theory and Its Application in Heat Engineering, Proceedings of the First Inter-Institutional Conference, No. 139 (Transzheldorizdat, 1961).

24. L. A. Goryainov, Analytical Methods for the Solution of Problems in the Heat Transfer between a Moving Emitting Medium and a Heat-Sensitive Surface, Tr. MIITa, No. 125 (Transzheldorizdat, 1960).

25. M. T. Smirnov, Heat Transfer in Gases by Radiation and Contact Simultaneously, Izv. VTI, No. 3 (1929).

26. V. T. Kumskov and L. A. Goryainov, Laws of Composite Heat Transfer, Tr. MIITa, No. 125 (Transzheldorizdat, 1960).

27. E. R. G. Eckert, Introduction to the Theory of Heat and Mass Transfer [Russian translation] (Gosénergoizdat, 1957).

28. P. K. Konakov, On the Regularities of Composite Heat Transfer, Intern. J. Heat Mass Transfer, Vol. 2, 136–149 (1961).

29. W. I. Wohlenberg and D. G. Morrow, Radiation in the Pulverized-Fuel Furnace, Trans. ASME, Vol. 47 (1925).

30. M. Adams, Heat Transmission, Vol. 3 (1954).

31. M. Ledinegg, Dampferzeugung [Steam Generation] (1952).

32. G. Ribaud and E. Brun, Transmission de la chaleur [Heat Transmission], 1 (1948).

33. A. Zinzen, Dampfkessel und Feuerungen [Boilers and Furnaces] (1950).

34. H. C. Hottel and R. T. Haslam, Combustion and Heat Transfer, Trans. ASME, Vols. 49 and 50 (1927–1928).

35. H. C. Hottel and H. G. Mangelsdorf, Heat Transmission from Non-Luminous Gases: II. Experimental Study of CO_2 and H_2O, Trans. Am. Inst. Chem. Engrs., Vol. 31, 517–549 (1935).

36. H. C. Hottel and R. B. Egbert, Trans. Am. Inst. Chem. Engrs., Vol. 38, 531 (1942).

37. S. S. Penner, The Emission of Radiation from Diatomic Cases, J. Appl. Phys., 21 (7):(1950); 22 (9):(1951); 23 (2):(1952); 28 (5):(1957).

38. G. N. Plass, Spectral Emissivity of Carbon Dioxide from 1800–2500 cm^{-1}, J. Opt. Soc. Am., 49 (9):(1959).

39. I. T. Bevans, R. V. Dunkle, D. E. Edwards, I. T. Giery, L. L. Levensen, and A. P. Oppenheim, Apparatus for the Determination of the Band Absorption of Gases at Elevated Pressures and Temperatures, J. Opt. Soc. Am., 50:130 (1960).

40. D. K. Edwards, Absorption of Infrared Bands of Carbon Dioxide at Elevated Pressures and Temperatures, J. Opt. Soc. Am., 50:617–666 (1960).

41. M. Czerny and L. Genzel, Glastech. Ber., 25:134–139 (1952); 25:387–393 (1952).

42. L. Genzel, Glastech. Ber., 26:69–71 (1953); Z. Physik, 135:177–195 (1953).

43. V. A. Walter, I. Dörr, and E. Eller, Glastech. Ber., 26:133–140 (1953).

44. M. Goulard and R. Goulard, Intern. J. Heat Mass Transfer, 1:81–91 (1960).

45. A. F. Sarofim, Sc. D. Thesis in Chemical Engineering, M. I. T., Cambridge, Mass. (in preparation).

46. C. M. Usiskin and E. M. Sparrow, Thermal Radiation between Parallel Plates Separated by an Absorbing-Emitting Nonisothermal Gas, Intern. J. Heat Mass Transfer, 1:(1960); Trans. ASME, Series C: J. Heat Transfer, 83 (2):(1961).

47. R. Viskanta and R. I. Grosh, Heat Transfer by Simultaneous Conduction and Radiation in an Absorbing Medium, Trans. ASME, Series C: J. Heat Transfer, 83:(1961); 84:(1962).

48. R. Viskanta and R. I. Grosh, Effect of Surface Emissivity on Heat Transfer by Simultaneous Conduction and Radiation, Intern. J. Heat Mass Transfer, 5:729–734 (1962).

49. R. Viskanta and R. I. Grosh, Boundary Layer in Thermal Radiation Absorbing and Emitting Media, Intern. J. Heat Mass Transfer, 5:795–806 (1962).

50. S. Rosseland, Astrophysics Based upon Atomic Theory [Russian translation] (ONTI, 1936).

51. K. Elgeti, Ein neues Verfahren zur Berechnung des Strahlungsaustausches zwischen einem Gas und einer grauen Wand [A New Method for Calculating the Radiative Exchange between a Gas and a Gray Wall], Brenstoff-Waerme-Kraft, 14 (1):1–40 (1962).

52. H. C. Hottel, Some Problems in Radiative Transport, presented at the 1961 International Heat Transfer Conference, August 28-September 1, 1961, University of Colorado, Boulder, Colorado, USA; Intern. J. Heat Mass Transfer, 5:82–83 (1962).

53. I. P. Kolchenogova and S. N. Shorin, Investigation of Radiant Energy Transfer in an Attenuating Medium, Izv. Akad. Nauk SSSR, Otd. Tekhn. Nauk, No. 5 (1956).

54. B. B. Galitsyn, On Radiant Energy, Uch. Zap. Mosk. Univ., Otd. Fiz.-Mat. Nauk (Scientific Writings of Moscow University, Department of Physicomathematical Sciences), No. 10 (1893).

55. V. N. Adrianov and G. L. Polyak, Differential Methods for Studying Radiant Heat Transfer, Intern. J. Heat Mass Transfer, 6 (5):355–362 (1963).

ISOTHERMAL MOTION OF A VISCOUS FLUID PAST A THIN PLATE

P. K. Konakov

We wish to examine the isothermal steady-state process of a fluid with constant density ρ and kinematic viscosity coefficient ν flowing past a thin plate. We will assume that the fluid free stream is parallel to the plate and has a small velocity w_0.

It is postulated in boundary layer theory that a boundary layer of variable thickness δ is formed about the plate, and that in this layer the lengthwise component of the flow velocity w_x varies from zero on the plate to the free stream value w_0, no mention being made in the meantime of the boundary layer associated with the transverse velocity component w_y. Nevertheless, the component w_y turns out to have a considerable influence on the motion of a viscous fluid past a plate, to the extent that it becomes convenient to assume that two boundary layers are effectively formed; in the first, of thickness δ, the component w_x varies from 0 to w_0; in the second, of thickness Δ, the component w_y varies from zero at the plate to zero in the free stream. It will be shown below that $\Delta > \delta$.

The notion of a second boundary layer gives a basis for analyzing with physical credence the problem of isothermal motion of a viscous fluid about a thin plate.

We will suppose that the boundary layer of thickness Δ formed on the plate is laminar. The equations for a two-dimensional laminar boundary layer of fluid without a pressure gradient have the form

$$\frac{\partial w_x}{\partial x} + \frac{\partial w_y}{\partial y} = 0, \tag{1}$$

$$w_x \frac{\partial w_x}{\partial x} + w_y \frac{\partial w_x}{\partial y} = \nu \frac{\partial^2 w_x}{\partial y^2}. \tag{2}$$

The conditions for uniqueness of the given problem are made up of geometric and boundary conditions. The geometric conditions for the plane of the plate in contact with the moving fluid are determined by the equation $y = 0$. The equation of surface contact between the second laminar boundary layer and the undisturbed stream has the form $\Delta = \varphi(x)$.

The boundary conditions are formulated as follows:

$$\left| w_x \right|_{y=0} = 0, \; \left| w_y \right|_{y=0} = 0, \; \left| w_x \right|_{y=\Delta} = w_0, \; \left| \frac{dw_x}{dy} \right|_{y=\Delta} = 0,$$
$$\left| w_y \right|_{y=\Delta} = 0.$$

It is assumed in boundary layer that the curves for the velocity w_x as a function of the coordinate y are similar to one another in the first boundary layer. Hence

$$w_x = w_0 \varphi(\eta),$$

where

$$\eta = \frac{y}{\delta}. \tag{3}$$

Differentiating both sides of Eq. (3) with respect to x, we obtain

$$\frac{\partial w_x}{\partial x} = - w_0 \varphi' \eta \frac{1}{\delta} \frac{d\delta}{dx}.$$

Taking into account the equation of continuity (1), we write

$$w_y = \int_0^y \frac{\partial w_x}{\partial y}\, dy = -\int_0^y \frac{\partial w_x}{\partial x}\, dy = w_0 \frac{d\delta}{dx} \int_0^\eta \varphi' \eta\, d\eta.$$

Since

$$\int_0^\eta \varphi' \eta\, d\eta = \varphi\eta - \int_0^\eta \varphi\, d\eta,$$

it follows that

$$w_y = w_0 \frac{d\delta}{dx}\left[\varphi\eta - \int_0^\eta \varphi\, d\eta\right]. \tag{4}$$

Equations (3) and (4) permit Eq. (2) to be rewritten in the form

$$-w_0^2 \varphi\varphi'\eta \frac{1}{\delta}\frac{d\delta}{dx} + w_0^2 \frac{d\delta}{dx}\left[\varphi\eta - \int_0^\eta \varphi\, d\eta\right]\varphi'\frac{1}{\delta} = \nu w_0 \varphi''\frac{1}{\delta^2}.$$

After some straightforward manipulations we obtain the following integrodifferential equation:

$$-\varphi'\int_0^\eta \varphi\, d\eta = \frac{\nu}{w_0\delta}\cdot\frac{1}{\dfrac{d\delta}{dx}}\,\varphi''. \tag{5}$$

This equation may be written in the form

$$-\int_0^\eta \varphi\, d\eta = A\frac{d\ln\varphi'}{d\eta},$$

$$A = \frac{\nu}{w_0\delta}\cdot\frac{1}{\dfrac{d\delta}{dx}},$$

whence

$$\varphi' = \varphi'_{\text{w}}\, e^{-\frac{1}{A}\int_0^\eta d\eta \int_0^\eta \varphi\, d\eta},$$

which yields

$$\varphi = \varphi'_{\text{w}}\int_0^\eta e^{-\frac{1}{A}\int_0^\eta d\eta \int_0^\eta \varphi\, d\eta}\, d\eta. \tag{6}$$

In these expressions φ'_{w} denotes the value of the function φ' in the plate. Let

$$\varphi = a_0 + a_1\eta + a_2\eta^2 + a_3\eta^3 + \cdots\,;$$

then substituting the function φ and its derivatives into Eq. (5), we obtain

$$-(a_1+2a_2\eta+3a_3\eta+\dots)\left(a_0\eta+\frac{a_1}{2}\eta^2+\frac{a_2}{3}\eta^3+\dots\right)=$$

$$= A(2a_2+6a_3\eta+12a_4\eta^2+\dots).$$

Performing series multiplication, we have

$$-\left(a_0a_1\eta+\frac{a_1^2}{2}\eta+\dots+2a_0a_2\eta^2+a_1a_2\eta^3+\dots+3a_0a_3\eta^3+\right.$$

$$\left.+\frac{3}{2}a_1a_3\eta^4+\dots\right) = A(2a_2+6a_3\eta+\dots).$$

From a comparison of the coefficients associated with terms of identical powers of η, we obtain the equations

$$a_2 = 0, \quad -a_0a_1 = 6Aa_3,$$

$$-\frac{a_1^2}{2}-2a_0a_2=12Aa_4, \quad \dots.$$

The boundary conditions give

$$a_0 = 0,$$

so that

$$a_3 = 0, \quad a_4 = \frac{a_1^2}{24A}.$$

Consequently,

$$\frac{w_x}{w_0} = a_1\eta - \frac{a_1^2}{24A}\eta^4 + \frac{11d^3}{5040A^2}\eta^7 - \dots.$$

Limiting ourselves to the first two terms of the series, it may be approximately stated that

$$\frac{w_x}{w_0} = a_1\eta - \frac{a_1^2}{24A}\eta^4. \tag{7}$$

With this relation, Eq. (4) is written as

$$\frac{w_y}{w_0} = \frac{d\delta}{dx}\left(\frac{a_1}{2}\eta^2 - \frac{a_1^2}{30A}\eta^5\right). \tag{8}$$

We denote the ratio Δ/δ by ξ. Equation (8) combined with the fifth boundary condition yields

$$\frac{a_1}{2}\xi^2 - \frac{a_1^2}{30A}\xi^5 = 0,$$

whence

$$\xi = \sqrt[3]{15\frac{A}{a_1}}.$$

We will assume that the curves for the variation of the velocity w_x with the coordinate y_Δ in the second boundary layer are similar to the analogous curves in the first boundary layer, i.e., that Eq. (7) is also valid for the secondary boundary layer. In this case the variable η of Eq. (7) is defined by the relation

$$\eta = \frac{y_\Delta}{\Delta} \quad \text{or} \quad \eta = \frac{y_\Delta}{\xi\delta} \; .$$

Substituting this value of η into Eq. (7), we obtain

$$\frac{w_x}{w_0} = \frac{a_1}{\xi} \eta - \frac{a_1^2}{24A\xi^4} \eta^4. \tag{9}$$

In this equation the variable η varies as in Eq. (8) from 0 to ξ. Substituting ξ into the right-hand side of Eq. (9), we obtain the following equation on the basis of the third boundary condition:

$$a_1 - \frac{a_1^2}{24A} = 1.$$

The fourth boundary condition gives

$$a_1 - \frac{a_1^2}{6A} = 0.$$

Solving the boundary equations, we have

$$a_1 = \frac{4}{3} , \quad A = \frac{2}{9} ,$$

hence

$$\frac{\nu}{w_0\delta} \cdot \frac{1}{\dfrac{d\delta}{dx}} = \frac{2}{9} \; .$$

The solution of this equation, in compliance with the boundary conditions, leads to the formula

$$\delta = 3 \sqrt{\frac{\nu x}{w_0}} \; . \tag{10}$$

Substituting the numerical values found for a_1 and A into the expression for ξ, we find

$$\xi = \sqrt[3]{15 \cdot \frac{2 \cdot 3}{9 \cdot 4}} = 1.357.$$

Now Eq. (9) can be written in the following form:

$$\frac{w_x}{w_0} = 0.982\eta - 0.0982\eta^4, \tag{11}$$

and, since

$$\frac{d\delta}{dx} = \frac{1.5}{\sqrt{\dfrac{w_0 x}{\nu}}} \; ,$$

we finally obtain

$$\frac{w_y}{w_0} = \frac{1}{\sqrt{\dfrac{w_0 x}{\nu}}} (\eta^2 - 0.4\eta^5). \tag{12}$$

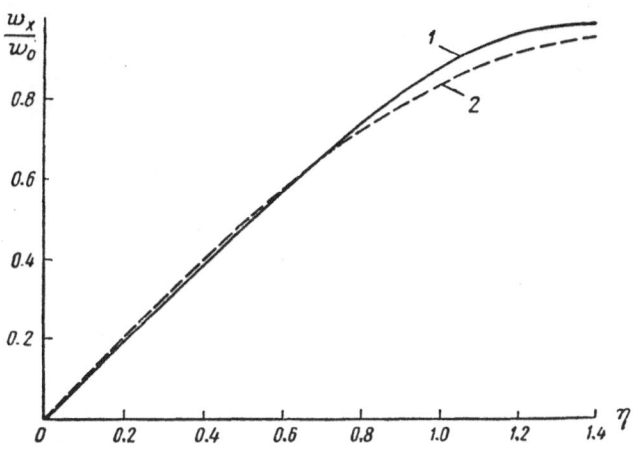

Graph of $wx/w_0 = f(\eta)$: 1) According to Eq. (11); 2) Blasius solution.

The figure presents two curves—one showing the dependence (11) and the other representing the familiar Blasius solution. Both curves essentially coincide. The following expression is valid for the relative value of the drag stress $S_W/\rho w_0^2$ on the plate:

$$\frac{S_W}{\rho w_0^2} = \frac{\nu}{w_0^2}\left|\frac{dw_x}{dy}\right|_W,$$

and, since

$$\left|\frac{dw_x}{dy}\right|_W = \frac{w_0 a_1}{\xi\delta} = \frac{w_0 a_1}{3\xi\sqrt{\dfrac{\nu x}{w_0}}},$$

we have

$$\frac{S_W}{\rho w_0^2} = \frac{a_1}{3\xi\sqrt{\dfrac{w_0 x}{\nu}}}.$$

Substituting into this equation the values found for a_1 and ξ, we obtain

$$\frac{S_W}{\rho w_0^2} = \frac{0.329}{\sqrt{\dfrac{w_0 x}{\nu}}}. \qquad \text{or} \qquad \frac{S_W}{\rho w_0^2} = \frac{0.329}{\sqrt{Re}}, \qquad (13)$$

which almost agrees with the well-known Blasius formula.

NONISOTHERMAL MOTION OF A VISCOUS FLUID PAST A THIN PLATE

P. K. Konakov

We now consider the nonisothermal steady-state flow of a fluid with constant density ρ and kinematic viscosity coefficient ν past a thin isothermal plate with temperature T_w.

We will suppose that the hydrodynamic boundary layer of thickness Δ formed on the plate is laminar. Then the equations of this layer are written as

$$\frac{\partial w_x}{\partial x} + \frac{\partial w_x}{\partial y} = 0, \tag{1}$$

$$w_x \frac{\partial w_x}{\partial x} + w_v \frac{\partial w_x}{\partial y} = \nu \frac{\partial^2 w_x}{\partial y^2}. \tag{2}$$

It is postulated in temperature boundary layer theory that a temperature boundary layer of variable thickness δ_T is formed about the plate, the temperature T in this layer varying from T_w at the plate to T_0 in the thermally unperturbed fluid flow.

By analogy with the hydrodynamic boundary layer, it is convenient to introduce the notion that two temperature boundary layers are formed about the plate. In the first layer, of thickness δ_T, the component $\rho c T w_x$ varies from 0 at the plate to a value $\rho c T_0 w_{x_0}$ at the outer surface of the layer; in the second layer, of thickness Δ, the component $\rho c T w_y$ varies from 0 at the plate to a value $\rho c T_0 w_{y_0}$ at the outer surface of the layer.

This concept permits a proper physical interpretation of the problem of nonisothermal motion of a viscous fluid past a thin plate.

We will assume that the temperature boundary layer of thickness Δ_T is situated within the limits of the hydrodynamic boundary layer of thickness Δ. Then the equation of the temperature boundary layer is formulated as

$$w_x \frac{\partial T}{\partial x} + w_y \frac{\partial T}{\partial y} = a \frac{\partial^2 T}{\partial y^2}, \tag{3}$$

where a is the coefficient of thermal conductivity.

To the boundary conditions formulated in the preceding section we add the equations

$$\left. T \right|_{y=0} = T_w, \left. T \right|_{y=\Delta} = T_0, \left. \frac{\partial T}{\partial y} \right|_{\Delta y = \tau} = 0.$$

It follows from an examination of Eqs. (2) and (3) that the curves for the variation of w_x and $T - T_w$ with the coordinate y are similar in the hydrodynamic and temperature boundary layers. If $\nu = a$ (Pr = 1), these curves are congruent. On the basis of this similarity we write the equation

$$\frac{\Delta_T}{\delta_T} = \xi = 1.357.$$

The following relations are applicable to the temperature boundary layer:

$$\frac{\partial T}{\partial x} = -\frac{dT}{d\eta_\tau}\,\eta_\tau\,\frac{1}{\delta_\tau}\cdot\frac{d\delta_\tau}{dx},$$

$$\frac{\partial T}{\partial y} = \frac{dT}{d\eta_\tau}\cdot\frac{1}{\delta_\tau},\quad \frac{\partial^2 T}{\partial y^2} = \frac{d^2 T}{d\eta^2}\cdot\frac{1}{\delta_\tau^2}.$$

Denoting the ratio δ_T/δ by ξ_T, we can write

$$\delta_\tau = \delta\xi_\tau.$$

Consequently,

$$\frac{\partial T}{\partial x} = -\frac{dT}{d\,(\xi_\tau\eta)}\cdot\xi_\tau\eta\,\frac{1}{\delta}\frac{d\delta}{dx},$$

$$\frac{\partial T}{\partial y} = \frac{dT}{d\,(\xi_\tau\eta)}\cdot\frac{1}{\delta},\quad \frac{\partial^2 T}{\partial y^2} = \frac{d^2 T}{d\,(\xi_\tau\eta)^2}\cdot\frac{1}{\delta_2}.$$

In these relations the independence variable is the product $\xi_T\eta$. With this variable in mind, Eq. (3) can be rewritten as

$$-\,w_0\varphi\,(\xi_\tau\eta)\frac{dT}{d\,(\xi_\tau\eta)}\cdot(\xi_\tau\eta)\,\frac{1}{\delta}\cdot\frac{d\delta}{dx}\,+$$

$$+\,w_0\frac{d\delta}{dx}\left[\varphi\,(\xi_\tau\eta)\,(\xi_\tau\eta) - \int_0^{\xi_\tau\eta}\varphi\,(\xi_\tau\eta)\,d\,(\xi_\tau\eta)\right]\times$$

or

$$\times\frac{dT}{d\,(\xi_\tau\eta)}\cdot\frac{1}{\delta} = a\,\frac{d^2 T}{d\,(\xi_\tau\eta)^2}\cdot\frac{1}{\delta^2}$$

$$-\frac{dT}{d\,(\xi_\tau\eta)}\int_0^{\xi_\tau\eta}\varphi\,(\xi_\tau\eta)\,d\,(\xi_\tau\eta) = \frac{a}{w_0\delta}\cdot\frac{1}{\dfrac{d\delta}{dx}}\cdot\frac{d^2 T}{d\,(\xi_\tau\eta)^2}, \tag{4}$$

where

$$\int_0^{\xi_\tau\eta}\varphi\,(\xi_\tau\eta)\,d\,(\xi_\tau\eta) = \frac{d_1}{2}\,(\xi_\tau\eta)^2 - \frac{a_1}{120A}\,(\xi_\tau\eta)^5. \tag{5}$$

We will seek the solution of Eq. (4) in series form:

$$\frac{T - T_W}{T_0 - T_W} = b_1\,(\xi_\tau\eta) + b_2\,(\xi_\tau\eta)^2 + \dots.$$

Determining the first and second derivatives of T with respect to $\xi_T\eta$ and taking Eq. (5) into account, we can write Eq. (4) as follows:

$$\left[-\frac{a_1}{2}\,(\xi_\tau\eta)^2 + \frac{a_1^2}{120A}\,(\xi_\tau\eta)^5\right]\cdot[b_1 + 2b_2\,(\xi_\tau\eta) + 3b_3\,(\xi_\tau\eta)^2 + \dots] =$$

$$= \frac{A}{\mathrm{Pr}}\,[2b_2 + 6b_3\,(\xi_\tau\eta) + 12b_4\,(\xi_\tau\eta)^2 + \dots]$$

or

$$-\frac{a_1 b_1}{2}(\xi_\tau\eta)^2 + \frac{a_1^2 b_1}{120A}(\xi_\tau\eta)^5 - a_1 b_2(\xi_\tau\eta)^3 + \frac{a_1^2 b_2}{60A}(\xi_\tau\eta)^6 - \frac{3}{2}a_1 b_2(\xi_\tau\eta)^4 +$$

$$+ \ldots = \frac{A}{\mathrm{Pr}}[2b_2 + 6b_3(\xi_\tau\eta) + 12b_4(\xi_\tau\eta)^2 + \ldots].$$

We see from a comparison of the coefficients of like powers of $\xi_T\eta$ that

$$b_2 = 0, \ b_3 = 0, \ b_4 = \frac{a_1 b_1 \, \mathrm{Pr}}{24A}, \ b_5 = 0,$$

these relations enable us to write the approximate relation

$$\frac{T - T_W}{T_0 - T_W} = b_1(\xi_\tau\eta) - \frac{a_1 b_1 \, \mathrm{Pr}}{24A}(\xi_\tau\eta)^4. \tag{6}$$

The boundary conditions yield the following equations:

$$b_1 - \frac{a_1 b_1 \xi_\tau^3 \, \mathrm{Pr}}{6A} = 0,$$

$$b_1 \xi_\tau - \frac{b_1 \xi_\tau a_1 \xi_\tau^3 \, \mathrm{Pr}}{24A} = 1.$$

Solving these equations, we find

$$6A = a_1 \xi_\tau^3 \, \mathrm{Pr},$$

$$b_1 \xi_\tau = \frac{4}{3} = a_1.$$

It was established in the preceding article that

$$6A = a_1,$$

hence

$$\xi_\tau^3 \, \mathrm{Pr} = 1,$$

so that

$$\xi_\tau = \frac{1}{\sqrt[3]{\mathrm{Pr}}}.$$

We will assume that the curves for the variation of the temperature difference $T - T_W$ with the coordinate y_{Δ_T} in the second temperature boundary layer are similar to the analogous curves in the first temperature boundary layer, i.e., Eq. (6) is also valid for the boundary layer of thickness Δ_T. In this case

$$\eta = \frac{y_{\Delta_T}}{\Delta_T} = \frac{y_{\Delta_T}}{\xi\delta_T},$$

and Eq. (6) is rewritten as

$$\frac{T - T_W}{T_0 - T_W} = \frac{a_1}{\xi_\tau\xi}\eta - \frac{a_1^2}{24A\xi_\tau^4\xi}\eta^4. \tag{7}$$

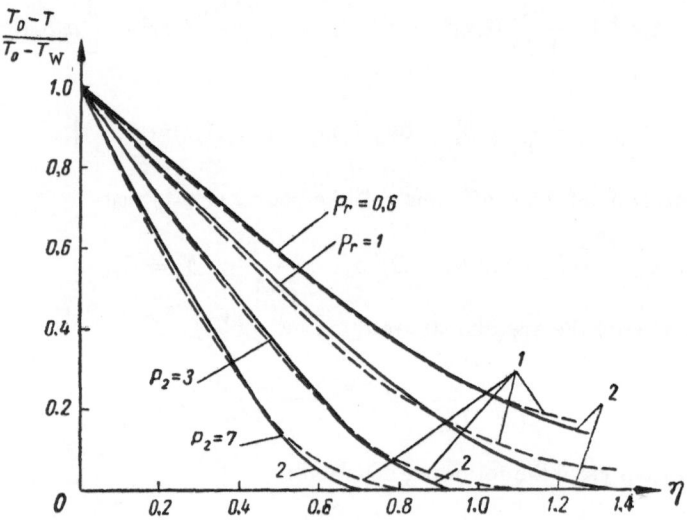

Graph of the dependence $(T_0-T)/(T_0-T_W) = f(\eta)$. 1) According
to Eq. (8); 2) Pohlhausen solution.

This relation is valid for the temperature boundary layer of thickness Δ_T, where η varies from 0 to $\xi_T\xi$. Making use of the numerical values of the quantities appearing on the right-hand side of the relation (7), we rewrite it in the following final form:

$$\frac{T-T_W}{T_0-T_W} = 0.982\sqrt[3]{\mathrm{Pr}}\,\eta - 0.0982\,\mathrm{Pr}^{\frac{4}{3}}\,\eta^4. \tag{8}$$

In the figure curves are shown which are constructed according to the dependence (8) and curves for the well-known Pohlhausen solution for different values of Pr. Both families of curves are practically congruent.

The specific heat flux q(x), of course, is defined by the equation

$$q(x) = \lambda\left|\frac{dT}{dy}\right|_{y=0}; \tag{9}$$

the minus sign has been dropped from the right-hand side of this equation because the direction of q(x) is opposite that of y.

Determining the derivative $|dT/dy|_{y=0}$ from Eq. (8) and substituting it into Eq. (9), and then taking into account the equation

$$\xi_T = \frac{1}{\sqrt[3]{\mathrm{Pr}}} \quad \text{and} \quad \delta = 3\sqrt{\frac{\nu x}{w_0}},$$

we obtain

$$q(x) = 0.329\frac{\lambda}{x}\sqrt[3]{\mathrm{Pr}}\,\sqrt{\mathrm{Re}}\,(t_0 - t_W), \tag{10}$$

which essentially coincides with the well-known Pohlhausen formula.

Let us now examine the nonisothermal steady-state process of a compressible fluid with constant dynamic viscosity μ and coefficient of thermal conductivity λ moving past an isothermal plate with a temperature T_W. Under laminar flow conditions in the boundary layers the defining system of equations has the form

$$\frac{\partial (\rho w_x)}{\partial x} + \frac{\partial (\rho w_y)}{\partial y} = 0,$$

(11)

$$\rho w_x \frac{\partial w_x}{\partial x} + \rho w_y \frac{\partial w_x}{\partial y} = \mu \frac{\partial^2 w_x}{\partial y^2},$$

(12)

$$\rho w_x \frac{\partial T}{\partial x} + \rho w_y \frac{\partial T}{\partial y} = \frac{\lambda}{c_p} \cdot \frac{\partial^2 T}{\partial y^2},$$

(13)

$$p = \rho R T.$$

(14)

It is necessary to add to this system of equations the above-mentioned uniqueness conditions.

We assume on the basis of the similarity principle that

$$\frac{\rho w_x}{\rho_0 w_0} = \varphi (\eta),$$

whence

$$\frac{\partial (\rho w_x)}{\partial x} = - \rho_0 w_0 \frac{d\varphi}{d\eta} \cdot \eta \cdot \frac{1}{\delta} \cdot \frac{d\delta}{dx},$$

which gives

$$\rho w y = \int_0^y \frac{\partial (\rho w_y)}{\partial y}\, dy = - \int_0^y \frac{\partial (\rho w_x)}{\partial x}\, dy = \rho_0 w_0 \frac{d\delta}{dx} \int_0^\eta \eta\, d\varphi.$$

Otherwise,

$$\frac{\rho w_y}{\rho_0 w_0} = \frac{d\delta}{dx} \left(\varphi \eta - \int_0^\eta \varphi\, d\eta \right).$$

(15)

Equations (15) and (14) make it possible to rewrite Eq. (12) in the form

$$- \frac{d(\varphi T)}{d\eta} \int_0^\eta \varphi\, d\eta = \frac{\mu}{\rho_0 w_0 \delta \frac{d\delta}{dx}} \cdot \frac{d^2 (\varphi T)}{d\eta^2}.$$

(16)

With the help of (15), Eq. (13) is represented as follows:

$$- \frac{dT}{d\eta} \int_0^\eta \varphi\, d\eta = \frac{\lambda}{\rho_0 w_0 c_p \delta \frac{d\delta}{dx}} \cdot \frac{d^2 T}{d\eta^2}.$$

(17)

For Pr = 1 it follows from Eqs. (16) and (17) that

$$\frac{\dfrac{d^2 T}{d\eta^2}}{\dfrac{dT}{d\eta}} = \frac{\dfrac{d^2 (\varphi T)}{d\eta^2}}{\dfrac{d (\varphi T)}{d\eta}} .$$

Integration of this equation leads to the equation

$$\varphi T = c_1 T + c_2.$$

Determining the constants c_1 and c_2 from the boundary conditions

$$|\varphi|_W = 0, \quad |\varphi|_\delta = 1,$$

we obtain

$$\varphi = \frac{T_0}{T} \cdot \frac{T - T_W}{T_0 - T_W}.$$

Since

$$\varphi = \frac{\rho w_x}{\rho_0 w_0} = \frac{T_0 w_x}{T w_0},$$

it follows that

$$\frac{w_x}{w_0} = \frac{T - T_W}{T_0 - T_W},$$

i.e., the curves for the variation of w_x and $T - T_W$ with the coordinate y coincide.

Introducing the notation

$$\frac{T - T_W}{T_0 - T_W} = \psi,$$

we can write

$$\varphi = \frac{\psi}{\dfrac{T}{T_0}} = \frac{\psi}{\psi \left(1 - \dfrac{T_W}{T_0}\right) + \dfrac{T_W}{T_0}}.$$

With these equations in mind, we obtain from Eq. (16)

$$-\frac{d\psi}{d\eta} \int_0^\eta \frac{\psi \, d\eta}{\psi \left(1 - \dfrac{T_W}{T_0}\right) + \dfrac{T_W}{T_0}} = A \, \frac{d^2\psi}{d\eta^2}, \tag{18}$$

where

$$A = \frac{\mu}{\rho_0 w_0 \delta \dfrac{d\delta}{dx}},$$

and, since

$$\frac{dT}{d\eta} = \frac{d\psi}{d\eta},$$

Eq. (18) can also be derived from Eq. (17). We expand the integrand of Eq. (18) in a series:

$$\frac{\psi}{\psi \left(1 - \dfrac{T_W}{T_0}\right) + \dfrac{T_W}{T_0}} = b_0 + b_1 \psi + b_2 \psi^2 + \cdots.$$

Limiting ourselves to the first two terms, we obtain from the boundary conditions

$$b_0 = 0; \quad b_1 = 1,$$

as a result of which Eq. (18) is rewritten as

$$-\frac{d\psi}{d\eta} \int_0^\eta \psi\, d\eta = A \frac{d^2\psi}{d\eta^2}\ , \tag{19}$$

which agrees with Eq. (4) for Pr = 1.   This means that Eq. (8) will be valid for the investigated problem with Pr = 1:

$$\frac{T - T_W}{T_0 - T_W} = 0.982\eta - 0.0982\eta^4.$$

The following relation becomes valid for the quantity δ:

$$\delta = 3 \sqrt{\frac{\mu x}{\rho_0 w_0}}\ .$$

Equation (19) will also be valid when the ratio

$$\frac{T_W}{T_0} \to 1.$$

A METHOD FOR SOLVING THE EQUATIONS OF A NONISOTHERMAL LAMINAR BOUNDARY LAYER

P. K. Konakov

The equations of a nonisothermal laminar boundary have been given in the preceding article, "Non-isothermal Motion of a Viscous Fluid Past a Thin Plate."

These equations can be brought to the following two equations by means of the transformations indicated in that article:

$$-\frac{d\,[\varphi\,(\eta)\,\theta]}{d\eta}\int_0^\eta \varphi\,(\eta)d\eta = A_1\frac{d^2\,[\varphi\,(\eta)\,\theta]}{d\eta^2}\,, \tag{1}$$

$$-\frac{d\theta}{d\,(\xi_\tau\eta)}\int_0^{\xi_\tau\eta} \varphi\,(\xi_\tau\eta)\,d\,(\xi_\tau\eta) = A_2\frac{d^2\theta}{d(\xi_\tau\eta)^2}\,, \tag{2}$$

where

$$\varphi\,(\eta) = \frac{\rho w_x}{\rho_0 w_0}\,,\ \ \theta = \frac{T}{T_0 - T_W}\,,$$

$$A_1 = \frac{\mu}{\rho_0 w_0\delta\,\dfrac{d\delta}{dx}}\,,\ \ A_2 = \frac{\lambda}{\rho_0 w_0 c_p\delta\,\dfrac{d\delta}{dx}}\,,$$

we note that

$$A_2 = \frac{A_1}{\mathrm{Pr}}\,.$$

We will seek the solution of Eqs. (1) and (2) in the form of infinite series:

$$\varphi\,(\boldsymbol{\eta}) = a_0 + a_1\eta + a_2\eta^2 + a_3\eta^3 + \ldots,$$

$$\theta = b_0 + b_1\xi_\tau\eta + b_2\xi_\tau^2\eta^2 + b_3\xi_\tau^3\boldsymbol{\eta}^3 + \ldots.$$

Multiplying the series, we obtain

$$\varphi\,(\eta)\theta = a_0 b_0 + a_0 b_1\xi_\tau\eta + a_0 b_2\xi_\tau^2\eta^2 + a_0 b_3\xi_\tau^3\eta^3 + a_0 b_4\xi_\tau^4\eta^4 + \ldots + a_1 b_0\eta +$$

$$+\, a_1 b_1\xi_\tau\eta^2 + a_1 b_2\xi_\tau^2\eta^3 + a_1 b_3\xi_\tau^3\eta^4 + \ldots + a_2 b_0\eta^2 + a_2 b_1\xi_\tau\eta^3 +$$

$$+\, a_2 b_2\xi_\tau^2\eta^4 + a_2 b_3\xi_\tau^3\eta^5 + \ldots + a_3 b_0\eta^3 + a_3 b_1\xi_\tau\eta^4 + a_3 b_2\xi_\tau^2\eta^5 +$$

$$+\, a_3 b_3\xi_\tau^3\eta^6 + \ldots + a_4 b_0\eta^4 + a_4 b_1\xi_\tau\eta^5 + a_4 b_2\xi_\tau^2\eta^6 + a_4 b_3\xi_\tau^3\eta^7 + \ldots =$$

$$=\, a_0 b_0 + (a_0 b_1\xi_\tau + a_1 b_0)\,\eta + (a_0 b_2\xi_\tau^2 + a_1 b_1\xi_\tau + a_2 b_0)\,\boldsymbol{\eta}^2 +$$

$$+\, (a_0 b_3\xi_\tau^3 + a_1 b_2\xi_\tau^2 + a_3 b_1\xi_\tau + a_3 b_0)\,\eta^3 + (a_0 b_4\xi_\tau^4 + a_1 b_3\xi_\tau^3 + a_2 b_2\xi_\tau^2 + a_3 b_1\xi_\tau + a_4 b_0)\,\eta^4 + \ldots.$$

25

Differentiating the product $\varphi(\eta)\theta$ with respect to η:

$$\frac{d[\varphi(\eta)\theta]}{d\eta} = (a_0b_1\xi_\tau + a_1b_0) + 2(a_0b_2\xi_\tau^2 + a_1b_1\xi_\tau + a_2b_2)\eta + 3(a_0b_3\xi_\tau^3 +$$

$$+ a_1b_2\xi_\tau^2 + a_2b_1\xi_\tau + a_3b_0)\,\eta^2 + 4(a_0b_4\xi_\tau^4 + a_1b_3\xi_\tau^3 + a_2b_2\xi_\tau^2 + a_3b_1\xi_\tau +$$

$$+ a_4b_0)\,\eta^3 + 5(a_0b_5\xi_\tau^5 + a_1b_4\xi_\tau^4 + a_2b_3\xi_\tau^3 + a_3b_2\xi_\tau^2 + a_4b_1\xi_\tau + a_5b_0)\,\eta^4 + \ldots$$

A second differentiation yields the following result:

$$\frac{d^2[\varphi(\eta)\theta]}{d\eta^2} = 2(a_0b_2\xi_\tau^2 + a_1b_1\xi_\tau + a_2b_0) + 6(a_0b_3\xi_\tau^3 + a_1b_2\xi_\tau^2 +$$

$$+ a_2b_1\xi_\tau + a_3b_0)\,\eta + 12(a_0b_4\xi_\tau^4 + a_1b_3\xi_\tau^3 + a_2b_2\xi_\tau^2 + a_3b_1\xi_\tau + a_4b_0)\,\eta^2 +$$

$$+ 20(a_0b_5\xi_\tau^5 + a_1b_4\xi_\tau^4 + a_2b_3\xi_\tau^3 + a_3b_2\xi_\tau^2 + a_4b_1\xi_\tau + a_5b_0)\,\eta^3 + \ldots$$

Clearly,

$$\int_0^\eta \varphi(\eta)\,d\eta = a_d\eta + \frac{a_1}{2}\eta^2 + \frac{a_2}{3}\eta^3 + \frac{a_3}{4}\eta^4 + \ldots$$

Taking the above expressions into account, we write Eq. (1) as follows:

$$-\left(c_0a_0\eta + c_0\frac{a_1}{2}\eta^2 + c_0\frac{a_2}{3}\eta^3 + c_0\frac{a_3}{4}\eta^4 + \ldots + 2c_1a_0\eta^2 +\right.$$

$$+ c_1a_1\eta^3 + \frac{2}{3}c_1a_2\eta^4 + \frac{c_1a_3}{2}\eta^5 + \ldots + 3c_2a_0\eta^3 + \frac{3}{2}c_2a_1\eta^4 +$$

$$+ c_2a_2\eta^5 + \frac{3}{4}c_2a_3\eta^6 + \ldots + 4c_3a_0\eta^4 + 2c_3a_1\eta^5 + \frac{4}{3}c_3a_2\eta^6 +$$

$$\left. + c_3a_3\eta^7 + \ldots\right) = 2A_1c_1 + 6A_1c_2\eta + 12A_1c_3\eta^2 + 20A_1c_4\eta^3 + 30A_1c_5\eta^4 + \ldots,$$

where

$$c_0 = a_0b_1\xi_\tau + a_1b_0,$$

$$c_1 = a_0b_1\xi_\tau^2 + a_1b_1\xi_\tau + a_2b_0,$$

$$c_2 = a_0b_3\xi_\tau^3 + a_1b_2\xi_\tau^2 + a_2b_1\xi_\tau + a_3b_0,$$

$$c_3 = a_0b_4\xi_\tau^4 + a_1b_3\xi_\tau^3 + a_2b_2\xi_\tau^2 + a_3b_1\xi_\tau + a_4b_0,$$

$$c_4 = a_0b_5\xi_\tau^5 + a_1b_4\xi_\tau^4 + a_2b_3\xi_\tau^3 + a_3b_2\xi_\tau^2 + a_4b_1\xi_\tau + a_5b_0,$$

$$\cdots\cdots\cdots\cdots\cdots\cdots\cdots\cdots\cdots$$

From a comparison of the coefficients of like powers of η we obtain the equations

$$c_1 = 0,$$

$$a_0c_0 + 6A_1c_2 = 0,$$

$$\frac{c_0a_1}{2} + 2c_1a_0 + 12A_1c_3 = 0,$$

$$\frac{c_0 a_2}{3} + c_1 a_1 + 3c_2 a_0 + 20A_1 c_4 = 0.$$

It follows from the boundary conditions that

$$a_0 = 0, \qquad b_0 = \frac{T_W}{T_0 - T_W}.$$

Consequently,

$$c_0 = \frac{T_W}{T_0 - T_W} a_1, \quad c_1 = a_1 b_1 \xi_T + a_2 \frac{T_W}{T_0 - T_W} = 0,$$

$$c_2 = a_1 b_2 \xi_T^2 + a_2 b_1 \xi_T + a_3 b_0 = 0,$$

$$\frac{T_W}{T_0 - T_W} \frac{a_1^2}{2} + 12A_1 (a_1 b_3 \xi_T^3 + a_2 b_2 \xi_T^2 + a_3 b_1 \xi_T + a_4 b_0) = 0,$$

$$\frac{T_W}{T_0 - T_W} \frac{a_1 a_2}{3} + 20A_1 (a_1 b_4 \xi_T^4 + a_2 b_3 \xi_T^3 + a_3 b_2 \xi_T^2 + a_4 b_1 \xi_T + a_5 b_0) = 0.$$

We write the following series:

$$\frac{d()}{d(\xi_T \eta_i)} = b_1 + 2b_2 (\xi_T \eta_i) + 3b_3 (\xi_T \eta_i)^2 + 4b_4 (\xi_T \eta_i)^3 + \ldots + \frac{d^2()}{d(\xi_T \eta_i)^2} =$$

$$= 2b_2 + 6b_3 (\xi_T \eta_i) + 12b_4 (\xi_T \eta_i)^2 + 20b_5 (\xi_T \eta_i)^3 + \ldots + \int_0^{\xi_T \eta_i} \varphi (\xi_T \eta_i) \, d(\xi_T \eta) =$$

$$= a_0 (\xi_T \eta_i) + \frac{a_1}{2} (\xi_T \eta_i)^2 + \frac{a_2}{3} (\xi_T \eta_i)^3 + \frac{a_3}{4} (\xi_T \eta_i)^4 + \ldots.$$

Equation (2) is written as follows with the help of these series:

$$- \Big(b_1 a_0 \xi_T \eta + \frac{b_1 a_1}{2} \xi_T^2 \eta^2 + \frac{b_1 a_2}{3} \xi_T^3 \eta^3 + \frac{b_1 a_3}{4} \xi_T^4 \eta^4 + \ldots +$$

$$+ 2b_2 a_0 \xi_T^2 \eta^2 + b_2 a_1 \xi_T^3 \eta^3 + \frac{2}{3} b_2 a_2 \xi_T^4 \eta^4 + \frac{b_2 a_3}{2} \xi_T^5 \eta^5 + \ldots +$$

$$+ 3b_3 a_0 \xi_T^3 \eta^3 + \frac{3}{2} b_3 a_1 \xi_T^4 \eta^4 + b_3 a_2 \xi_T^5 \eta^5 + \frac{3}{4} b_3 a_3 \xi_T^6 \eta^6 + \ldots +$$

$$+ 4b_4 a_0 \xi_T^4 \eta^4 + 2b_4 a_1 \xi_T^5 \eta^5 + \frac{4}{3} b_4 a_2 \xi_T^6 \eta^6 + b_4 a_3 \xi_T^7 \eta^7 + \ldots \Big) =$$

$$= 2A_2 b_2 + 6A_3 b_3 \xi_T \eta + 12A_2 b_4 \xi_T^2 \eta^2 + 20A_2 b_5 \xi_T^3 \eta^3 + \ldots.$$

From a comparison of the coefficients of like powers of η we obtain the equations

$$b_0 = 0, \quad b_1 a_0 \xi_T + 6A_2 b_3 \xi_T = 0,$$

$$\frac{b_1 a_1}{2} \xi_T^2 + 2b_2 a_0 \xi_T^2 + 12A_2 b_4 \xi_T^2 = 0,$$

$$\frac{b_1 a_2}{3} \xi_T^3 + b_2 a_1 \xi_T^3 + 3b_3 a_0 \xi_T^3 + 20A_2 b_5 \xi_T^3 = 0,$$

and, since $a_0 = 0$, we have

$$b_3 = 0, \quad b_4 = -\frac{a_1 b_1}{24 A_2}.$$

With the relations $b_2 = 0$ and $b_3 = 0$, we obtain

$$a_2 = \frac{a_1 b_1 \xi_T}{\dfrac{T_W}{T_0 - T_W}}, \quad a_3 = \frac{a_1 b_1^2 \xi_T^2}{\left(\dfrac{T_W}{T_0 - T_W}\right)^2},$$

$$a_4 = -\frac{a_1 b_1^3 \xi_T^3}{\left(\dfrac{T_W}{T_0 - T_W}\right)^3} - \frac{a_1^2}{24 A_1}.$$

We are now in a position to write the following approximate relations:

$$\frac{\rho w_x}{\rho_0 w_0} = a_1 \eta - \frac{a_1 b_1 \xi_T}{\dfrac{T_W}{T_0 - T_W}} \eta^2 + \frac{a_1 b_1^2 \xi_T^2}{\left(\dfrac{T_W}{T_0 - T_W}\right)^2} \eta^3 -$$

$$- \frac{a_1 b_1^3 \xi_T^3}{\left(\dfrac{T_W}{T_0 - T_W}\right)^3} \eta^4 - \frac{a_1^2}{24 A_1} \eta^4, \tag{3}$$

$$\frac{T - T_W}{T_0 - T_W} = b_1 \xi_T \eta - \frac{a_1 b_1 \xi_T^4 \, \mathrm{Pr}}{24 A_1} \eta^4. \tag{4}$$

It follows from the boundary conditions that

$$\left| \frac{T - T_W}{T_0 - T_W} \right|_{\eta=1} = 1; \quad \left| \frac{dT}{d\eta} \right|_{\eta=1} = 0,$$

which gives

$$b_1 \xi_T - \frac{a_1 b_1 \xi_T^4 \, \mathrm{Pr}}{24 A_1} = 1, \tag{5}$$

$$b_1 \xi_T - \frac{a_1 b_1 \xi_T^4 \, \mathrm{Pr}}{6 A_1} = 0, \tag{6}$$

whence

$$b_1 \xi_T = \frac{4}{3}.$$

With this equation in mind, we rewrite the relation (3) in the form

$$\frac{\rho w_x}{\rho_0 w_0} = a_1 \eta - a_1 \vartheta_0 \eta^2 + a_1 \vartheta_0^2 \eta^3 - a_1 \vartheta_0^3 \eta^4 - \frac{a_1^2}{24 A_1} \eta^4,$$

where

$$\vartheta_0 = \frac{4\,(T_0 - T_W)}{3T_W}\,.$$

The boundary conditions

$$\left|\frac{\rho w_x}{\rho_0 w_0}\right|_{\eta=1} = 1, \quad \left|\frac{d\rho w_x}{d\eta}\right|_{\eta=1} = 0$$

yield the equations

$$a_1\left(1 - \vartheta_0 + \vartheta_0^2 - \vartheta_0^3\right) - \frac{a_1^2}{24A_1} = 1, \tag{8}$$

$$a_1\left(1 - 2\vartheta_0 + 3\vartheta_0^2 - 4\vartheta_0^3\right) - \frac{a_1^2}{6A_1} = 0, \tag{9}$$

so that

$$a_1 = \frac{4}{3 - 2\vartheta_0 + \vartheta^2}\,. \tag{10}$$

It follows from Eqs. (6) and (8) that

$$\frac{a_1 \xi_T^3 \,\mathrm{Pr}}{6A_1} = 1, \quad \frac{a_1}{6A_1} = 1 - 2\vartheta_0 + 3\vartheta_0^2 - 4\vartheta_0^3, \tag{11}$$

which gives

$$\xi_T = \frac{1}{\sqrt[3]{\left(1 - 2\vartheta_0 + 3\vartheta_0^2 - 4\vartheta_0^3\right)\mathrm{Pr}}}\,. \tag{12}$$

Equations (9) and (10) enable us to write down the equation

$$A_1 = \frac{2}{3\left(3 - 2\vartheta_0 + \vartheta_0^2\right)\left(1 - 2\vartheta_0 + 3\vartheta_0^2 - 4\vartheta_0^3\right)}\,. \tag{13}$$

Since

$$A_1 = \frac{\mu}{\rho_0 w_0 \delta \dfrac{d\delta}{dx}}\,,$$

it follows that

$$\delta = \sqrt{\frac{\mu x}{\rho_0 w_0}\,3\,(3 - 2\vartheta_0 + \vartheta_0^2)\,(1 - 2\vartheta_0 + 3\vartheta_0^2 - 4\vartheta_0^3)}\,. \tag{14}$$

The relations (4) and (7) are finally written in the form

$$\frac{\rho w_x}{\rho_0 w_0} = \frac{a_1}{\xi}\,\eta - \frac{a_1\vartheta_0}{\xi^2}\,\eta^2 + \frac{a_1\vartheta_0^2}{\xi^3}\,\eta^3 - \frac{a_1\vartheta_0^3}{\xi^4}\,\eta^4 - \frac{a_1^2}{24A_1\xi^4}\,\eta^4; \tag{15}$$

$$\frac{T - T_W}{T_0 - T_W} = \frac{4}{3\xi_T\xi}\,\eta_T - \frac{a_1\mathrm{Pr}}{18A_1\xi_T\xi^4}\,\eta_T^4. \tag{16}$$

In the resultant expression (15) the variable η varies from 0 to 1.357; in the expression (16) the variable η_T varies from 0 to $1.357 \, \xi$.

If ν is small the above relations transform to the following:

$$\frac{\rho w_x}{\rho_0 w_0} = 0.982\eta - 0.0982\eta^4,$$

$$\frac{T - T_W}{T_0 - T_W} = 0.982 \sqrt[3]{Pr}\eta_T - 0.0982 \, Pr^{4/3} \, \eta_T^4,$$

$$\xi_T = \frac{1}{\sqrt[3]{Pr}}, \quad \delta = 3\sqrt{\frac{\mu x}{\rho_0 w_0}}.$$

These approximate relations then comprise the solution of the stated problem.

SOLUTION OF THE COMPOSITE HEAT TRANSFER PROBLEM IN A MOVING GRAY MEDIUM WITH A SMALL OPTICAL DENSITY, ON THE BASIS OF THE BOUNDARY LAYER EQUATIONS

P. K. Konakov, V. T. Kumskov, Yu. P. Sidorov, and V. S. Sidorov

Considerable significance attaches to the problems of composite heat transfer in a moving gray medium. The combination transport of energy in gray media has not been treated in the literature to date. In the present article a solution is obtained on the basis of the boundary layer equations for the problem of composite heat transfer in a gray medium with low optical density.

We will consider the composite transfer of energy from a gray medium to a plate. Let a medium with density ρ, kinematic viscosity ν, and temperature T_0 flow along a plate at a velocity w_0. It is known that when this happens a boundary layer is formed near the surface of the plate. Let the wall temperature be equal to T_w. Let us suppose that the resultant boundary layer is laminar. The nonisothermal motion of the fluid is described by the following system of boundary layer equations:

$$\frac{\partial w_x}{\partial x} + \frac{\partial w_y}{\partial y} = 0,$$

$$w_x \frac{\partial w_x}{\partial x} + w_y \frac{\partial w_x}{\partial y} = \nu \frac{\partial^2 w_x}{\partial y^2},$$

$$w_x \frac{\partial T}{\partial x} + w_y \frac{\partial T}{\partial y} = a \frac{\partial^2 T}{\partial y^2}.$$

The solution of this system of equations was obtained in the preceding article by Konakov in the following form:

$$\frac{T - T_w}{T_0 - T_w} = 0.982 \sqrt[3]{\Pr \eta_\tau} - 0.0982 \Pr^{\frac{4}{3}} \eta_\tau^4 \tag{1}$$

$$T = T_w + 0.982(T_0 - T_w) \Pr^{\frac{1}{3}} \eta_\tau - 0.0982 \Pr^{\frac{4}{3}} \eta_\tau^4. \tag{2}$$

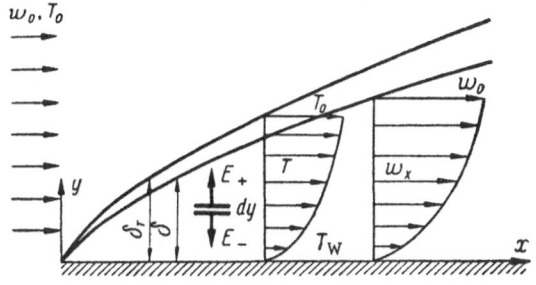

Fig. 1. Diagram of radiant energy transfer in a laminar boundary layer.

This solution describes the temperature field in the boundary layer in the convective transfer of thermal energy. We will obtain a solution for the resultant energy flux in the boundary layer for an absorbing-emitting fluid medium with absorption coefficient k, for the case of composite heat transfer. The transfer of radiant energy may be represented as two streams (Fig. 1). We denote the radiant energy flux away from the wall by E_+, the radiant energy flux toward the wall by E_-.

The equation for radiant energy transfer in the flow boundary layer is written in the form

$$\frac{dE_+}{d\eta_{\mathrm{T}}} + \kappa\delta E_+ = \kappa\delta\sigma_0 T^4, \tag{3}$$

$$- \frac{dE_-}{d\eta_{\mathrm{T}}} + \kappa\delta E_- = \kappa\delta\sigma_0 T^4, \tag{4}$$

where $\eta_{\mathrm{T}} = y/\delta_{\mathrm{T}}$ is the ratio of the coordinate y to the temperature thickness δ_{T} of the boundary layer, δ is the thickness of the hydrodynamic boundary layer in the x direction, σ_0 is the fourth-power law coefficient or black body radiation constant.

Equations (3) and (4) represent linear differential equations, which can be solved by the method of substitution. Making use of this method, the given equations are written in the form

$$u\frac{dz}{d\eta_{\mathrm{T}}} + \left(\frac{du}{d\eta_{\mathrm{T}}} + \kappa\delta u\right)z = \kappa\delta\sigma_0 T^4, \tag{5}$$

$$- u\frac{dz}{d\eta_{\mathrm{T}}} - \left(\frac{du}{d\eta_{\mathrm{T}}} - \kappa\delta u\right)z = \kappa\delta\sigma_0 T^4. \tag{6}$$

The solution of the equations are presented in the form

$$E_+ = e^{-\kappa\delta\eta_{\mathrm{T}}}\left(\kappa\delta\sigma_0 \int T^4 e^{\kappa\delta\eta_{\mathrm{T}}} d\eta_{\mathrm{T}} + c_1\right), \tag{7}$$

$$E_- = e^{\kappa\delta\eta_{\mathrm{T}}}\left(-\kappa\delta\sigma_0 \int T^4 e^{-\kappa\delta\eta_{\mathrm{T}}} d\eta_{\mathrm{T}} + c_2\right). \tag{8}$$

Utilizing Eq. (2), the expressions (7) and (8) are written

$$E_+ = \sigma_0\left[T_{\mathrm{w}}^4 - \frac{4aT_{\mathrm{w}}^3}{\kappa\delta} + \frac{4aT_{\mathrm{w}}^3\,\eta_{\mathrm{T}}}{1} + 6a^2 T_{\mathrm{w}}^2\eta_{\mathrm{T}}^2 - \right.$$
$$\left. - \frac{12a^2 T_{\mathrm{w}}\eta_{\mathrm{T}}}{\kappa\delta} + \frac{12a^2 T_{\mathrm{w}}^2}{(\kappa\delta)^2} + 4a^3 T_{\mathrm{w}}\eta_{\mathrm{T}}^3 - \frac{12a^3 T_{\mathrm{w}}\eta_{\mathrm{T}}^2}{\kappa\delta} + \frac{24a^3 T_{\mathrm{w}}\,\eta_{\mathrm{T}}}{\kappa\delta} - \frac{24a^3 T_{\mathrm{w}}}{(\kappa\delta)^3} \right] + c_1 e^{-\kappa\delta\eta_{\mathrm{T}}}. \tag{9}$$

$$E_- = \sigma_0\left[T_{\mathrm{w}}^4 + 4aT_{\mathrm{w}}^3\left(\eta_{\mathrm{T}} + \frac{1}{\kappa\delta}\right) + 6a^2 T_{\mathrm{w}}^2\left(\eta_{\mathrm{T}}^2 + \frac{2\eta_{\mathrm{T}}}{\kappa\delta} + \frac{2}{(\kappa\delta)^2}\right) + 4a^3 T_{\mathrm{w}}\left(\eta_{\mathrm{T}}^3 + \frac{3\eta_{\mathrm{T}}^2}{\kappa\delta} + \frac{6\eta_{\mathrm{T}}}{(\kappa\delta)^2} + \frac{6}{(\kappa\delta)^3}\right) \right] +$$
$$+ \left(a^4 - 4bT_{\mathrm{w}}^3\right)\left(\eta_{\mathrm{T}}^4 + \frac{4\eta_{\mathrm{T}}^3}{\kappa\delta} + \frac{12\eta_{\mathrm{T}}^2}{(\kappa\delta)^2} + \frac{24\eta_{\mathrm{T}}}{(\kappa\delta)^3} + \frac{24}{(\kappa\delta)^4}\right) \right] + c_2 e^{\kappa\delta\eta_{\mathrm{T}}}, \tag{10}$$

where

$$a = 0.982\,(T_0 - T_{\mathrm{w}})\sqrt[3]{\mathrm{Pr}}\,,$$

$$b = 0.0982\,(T_0 - T_{\mathrm{w}})\,\mathrm{Pr}^{\frac{4}{3}}\,.$$

Let us analyze the expression for the energy flux E_- from the medium to the wall.

In order to determine the integration constant we utilize the following boundary conditions. Suppose that the boundary layer extends sufficiently far in the direction of the coordinate y, in which case we may assume complete black body radiation at the boundary of the boundary layer. Then

$$\left| E_- \right|_{\eta_{\mathrm{T}}=1.357} = \sigma_0 T_0^4.$$

We determine c_2 from the boundary conditions

$$c_2 = e^{-1.357\kappa\delta}\sigma_0\left[\left(T_0^4 - T_W^4\right) - 4aT_W^3\left(1.357 + \frac{1}{\kappa\delta}\right) - \right.$$

$$\left. - 6a^2T_W^2\left(1.84 + \frac{2.714}{\kappa\delta} + \frac{2.0}{(\kappa\delta)^2}\right) - 4a^3T_W\left(2.49 + \frac{5.52}{\kappa\delta} + \frac{8.14}{(\kappa\delta)^2} + \frac{6}{(\kappa\delta)^3}\right)\right]. \tag{11}$$

We determine E_- with allowance for Eq. (11):

$$E_- = e^{\kappa\delta(\eta_T - 1.357)}\sigma_0\left[\left(T_0^4 - T_W^4\right) - 4aT_W^3\left(1.357 + \frac{1}{\kappa\delta}\right) - 6a^2T_W^2\left(1.84 + \frac{2.714}{\kappa\delta} + \frac{2}{(\kappa\delta)^2}\right)\right.$$

$$\left. - 4a^3T_W\left(2.49 + \frac{5.52}{\kappa\delta} + \frac{8.14}{(\kappa\delta)^2} + \frac{6}{(\kappa\delta)^3}\right)\right]$$

$$+ \sigma_0\left[T_W^4 + \frac{4aT_W^3}{\kappa\delta} + \frac{4aT_W^3\eta_T}{1} + 6a^2T_W^2\eta_T^2 + \frac{12a^2T_W^2\eta_T}{\kappa\delta} + \right.$$

$$\left. + \frac{12a^2T_W^2}{(\kappa\delta)^2} + \frac{4a^3T_W\eta_T^3}{1} + \frac{12a^3T_W\eta_T^2}{\kappa\delta} + \frac{24a^3T_W\eta_T}{(\kappa\delta)^2} + \frac{24a^3T_W}{(\kappa\delta)^3}\right]. \tag{12}$$

The resultant radiant energy flux at the wall may then be written

$$Q_{\scriptscriptstyle\Gamma} = |E_-|_{\eta_T=0} - |E_+|_{\eta_T=0}. \tag{13}$$

Making use of Eq. (12), we can find an expression for E_- at the wall when $\eta_T = 0$:

$$|E_-|_{\eta_T=0} = \sigma_0T_W^4 + \frac{4aT_W^3\sigma_0}{\kappa\delta} + \frac{12a^2T_W^2\sigma_0}{(\kappa\delta)^2} + \frac{24a^3T_W\sigma_0}{(\kappa\delta)^3} +$$

$$+ \frac{24(a^4 - 4bT_W^3\sigma_0)}{(\kappa\delta)^4} - e^{-1.357\kappa\delta}\sigma_0\left[\left(T_0^4 - T_W^4\right) + 4aT_W^3\left(1.357 + \frac{1}{\kappa\delta}\right) + \right.$$

$$+ 6a^2T_W^2\left(1.357^2 + \frac{2.714}{\kappa\delta} + \frac{2}{(\kappa\delta)^2}\right) + 4a^3T_W\left(1.357^3 + \frac{3\cdot 1.357^2}{\kappa\delta} + \frac{6\cdot 1.357}{(\kappa\delta)^2} + \frac{6}{(\kappa\delta)^3}\right) +$$

$$+ \left(a^4 - 4bT_W^2\right)\left(1.357^4 + \frac{4\cdot 1.357^3}{\kappa\delta} + \frac{12\cdot 1.357^2}{(\kappa\delta)^2} + \frac{24\cdot 1.357}{(\kappa\delta)^3} + \frac{24}{(\kappa\delta)^4}\right). \tag{14}$$

TABLE 1

$\dfrac{T_W}{T_0}$	$\kappa\delta$				
	1.0	0.9	0.7	0.5	0.3
0.9	0.211	0.222	0.246	0.262	0.299
0.8	0.354	0.37	0.409	0.444	0.506
0.7	0.446	0.47	0.521	0.573	0.651
0.6	0.502	0.527	0.588	0.659	0.748
0.5	0.532	0.56	0.623	0.713	0.803
0.3	0.550	0.573	0.634	0.762	0.856
0.1	0.561	0.582	0.644	0.771	0.869

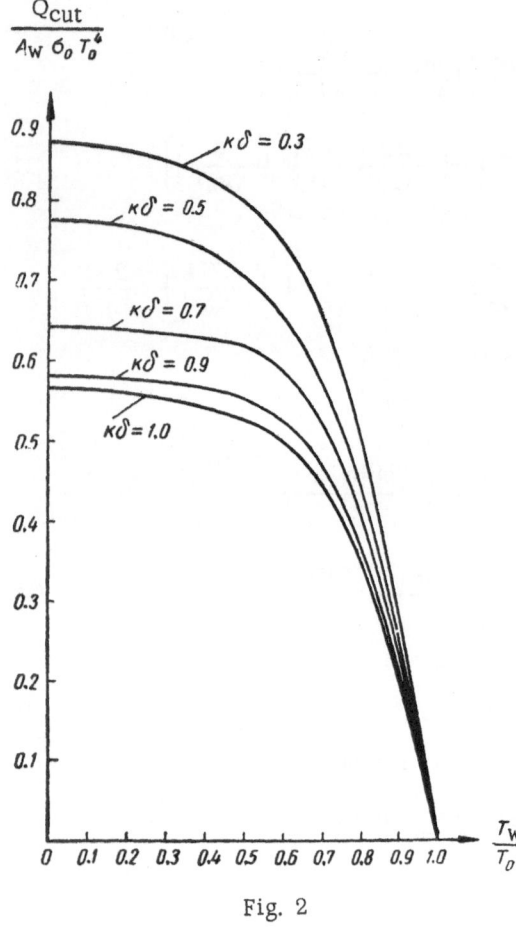

Fig. 2

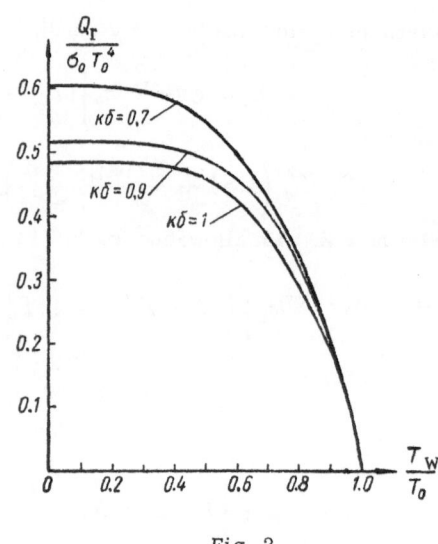

Fig. 3

From inspection of the boundary condition at the wall we can represent the expression for the flux E_+ in the form

$$|E_+|_{\eta_T=0} = |E_-|_{\eta_T=0}(1-A_W) + A_W \sigma_0 T_W^4, \quad (15)$$

where A_W is the absorptivity of the wall.

Substituting Eqs. (14) and (15) into (13), we obtain an expression for the resultant radiant flux in the gray medium at the wall:

$$Q_r = |E_-|_{\eta_T=0} - |E_+|_{\eta_T=0} = A_W \sigma_0 \left\{ e^{-1.357\kappa\delta}(T_0^4 - T_W^4) - 4aT_W^3\left[e^{-1.357\kappa\delta}\left(1.357 + \frac{1}{\kappa\delta}\right) - \frac{1}{\kappa\delta}\right] - \right.$$

$$- 6a^2 T_W^2 \cdot \left[e^{1.357\kappa\delta}\left(1.357^2 + \frac{2\cdot1.357}{\kappa\delta} + \frac{2}{(\kappa\delta)^2}\right) - \frac{2}{(\kappa\delta)^2}\right] -$$

$$\quad (16)$$

$$- 4a^3 T_W \cdot \left[e^{-1.357\kappa\delta}\left(1.357^3 + \frac{3\cdot1.357^2}{\kappa\delta} + \frac{6\cdot1.357}{(\kappa\delta)^2} + \frac{6}{(\kappa\delta)^3}\right) - \frac{6}{(\kappa\delta)^3}\right] -$$

$$\left. - (a^4 - 4bT_W^3) \cdot \left[e^{-1.357\kappa\delta}\left(1.357^4 + \frac{4\cdot1.357^3}{\kappa\delta} + \frac{12\cdot1.357^2}{(\kappa\delta)^2} + \frac{24\cdot1.357}{(\kappa\delta)^3} + \frac{24}{(\kappa\delta)^4}\right) - \frac{24}{(\kappa\delta)^4}\right] \right\}.$$

Figure 2 illustrates the dependence

$$\frac{Q_r}{A_W \sigma_0 T_0^4} = \varphi\left(\frac{T_W}{T_0}\right)$$

for different values of the optical density $k\delta$ of the medium, with Pr = 1. The data for the calculations are shown in Table 1.

TABLE 2

T_W	$\kappa\delta$		
T_0	0.7	0.9	1.0
0.9	0.22	0.21	0.19
0.8	0.345	0.315	0.29
0.6	0.495	0.46	0.425
0.5	0.55	0.49	0.46
0.3	0.595	0.512	0.48
0.1	0.602	1.515	0.482

Substituting Eq. (1) into the relation $Q_k = -\lambda\, dt/d\eta_T$, we obtain an expression for the convective component for $\eta_T = 0$:

$$Q_\kappa = 0.329 \sqrt[3]{\mathrm{Pr}}\, \sqrt{\mathrm{Re}}\, (T_0 - T_w). \qquad (17)$$

Once we know the expressions for Q_k and Q_r, we can write the total heat flux as

$$Q = Q_\kappa + Q_r. \qquad (18)$$

Utilizing the system of boundary layer equations, we can derive by analogy the expression for the resultant radiant flux at the boundary of the boundary layer. In this case we use the boundary condition

$$\left. E_- \right|_{\eta_T=1.357} = \sigma_0 T_0^4.$$

To evaluate the integration constant we equate (7) to the expression

$$E_+| = (1 - A_w)E_- + A_w \sigma_0 T_w^4, \qquad (19)$$

whereupon

$$c_1 = \left\{ \sigma_0 \left[T_w^4 + \frac{4aT_w^3}{\kappa\delta} + \frac{12a^2T_w^2}{(\kappa\delta)^2} + \frac{24a^3T_w}{(\kappa\delta)^3} + \frac{24\,(a^4 - 4bT_w^3)}{(\kappa\delta)^4} \right] + \right.$$

$$+ e^{-1.357\kappa\delta}\sigma_0 \left[(T_0^4 - T_w^4) - 4aT_w^3 \left(1.357 + \frac{1}{\kappa\delta} \right) - 6a^2T_w^2 \left(1.357^2 + \frac{2\cdot1.357}{\kappa\delta} + \frac{2}{(\kappa\delta)^2} \right) - \right.$$

$$- 4a^3T_w \left(1.357^3 + \frac{3\cdot1.357^2}{\kappa\delta} + \frac{6\cdot1.357}{(\kappa\delta)^2} + \frac{6}{(\kappa\delta)^3} \right) - \qquad (20)$$

$$- (a^4 - 4bT_w^3) \left(1.357^4 + \frac{4\cdot1.357^3}{\kappa\delta} + \frac{12\cdot1.357^2}{(\kappa\delta)^2} + \frac{24\cdot1.357}{(\kappa\delta)^3} + \frac{24}{(\kappa\delta^4)} \right) \right] \right\} \cdot (1 - A_w) +$$

$$+ A_w \sigma_0 T_w^4 - \sigma_0 \left[T_w^4 - \frac{4aT_w^3}{\kappa\delta} + \frac{12a^2T_w^2}{(\kappa\delta)^2} - \frac{24a^2T_w}{(\kappa\delta)^3} + \left(a^4 - 4bT_w^3 \right) \frac{24}{(\kappa\delta)^4} \right].$$

According to the expression (13), the resultant radiant energy flux at the boundary of the boundary layer for $\eta_T = 1.357$ is equal to

$$\left. |Q_r \right|_{\eta_T=1.357} = \sigma_0 T_0^4 - \sigma_0 \left[T_w^4 - \frac{4aT_w^3}{\kappa\delta} + 4aT_w^3 \cdot 1.357 + 6a^2T_w^2 \times 1.357^2 - \frac{12a^2T_w^2 \cdot 1.357}{\kappa\delta} + \right.$$

$$+ \frac{12a^2T_w^2}{(\kappa\delta)^2} + 4a^3T_w \cdot 1.357^3 - \frac{12a^3T_w\, 1.357^2}{\kappa\delta} + \frac{24a^3T_w\, 1.357}{\kappa\delta} - \frac{24a^3T_w}{(\kappa\delta)^3} + (a^4 - 4bT_w^3) \times$$

$$\times \left(1.357^4 - \frac{4\cdot1.357^3}{(\kappa\delta)} + \frac{12\cdot1.357^2}{(\kappa\delta)^2} - \frac{24\cdot1.357}{(\kappa\delta)^3} + \frac{24}{(\kappa\delta)^4} \right) \right] -$$

$$- e^{-1.357\kappa\delta}\sigma_0 \left\{ \left[T_w^4 + \frac{4aT_w^3}{\kappa\delta} + \frac{12a^2T_w^2}{(\kappa\delta)^2} + \frac{24a^3T_w}{(\kappa\delta)^3} + \frac{24\,(a^4 - 4bT_w^3)}{(\kappa\delta)^4} \right] \cdot (1 - A_w) + \right.$$

$$+ e^{-1.357\kappa\delta}\left[\left(T_0^4 - T_w^4\right) - 4aT_w^3\left(1.357 + \frac{1}{\kappa\delta}\right) - 6a^2T_w^2\left(1.357^2 + \frac{2\cdot1.357}{(\kappa\delta)} + \frac{2}{(\kappa\delta)^2}\right) - \right.$$

$$- 4a^3T_w\left(1.357^3 + \frac{3\cdot1.357^2}{\kappa\delta} + \frac{6\cdot1.357}{(\kappa\delta)^2} + \frac{6}{\kappa\delta^3}\right) +$$

$$+ (a^4 - 4bT_w^3)\left(1.357^4 + \frac{4\cdot1.357^3}{\kappa\delta} + \frac{12\cdot1.357^2}{(\kappa\delta)^2} + \frac{24\cdot1.357}{(\kappa\delta)^3} + \frac{24}{(\kappa\delta)^4}\right)\right](1 - A_w) +$$

$$+ A_w T_w^4 - \left[T_w^4 - \frac{4aT_w^3}{\kappa\delta} + \frac{12a^2T_w^2}{(\kappa\delta)^2} - \frac{24a^3T_w}{(\kappa\delta)^3} + (a^4 - 4bT_w^3)\frac{24}{(\kappa\delta)^4}\right]\right\}. \qquad (21)$$

Figure 3 presents a graph of the dependence

$$\frac{Q_r}{\sigma_0 T_0^4} = \psi\left(\frac{T_w}{T_0}\right)$$

for Pr = 1 and A_w = 1, for different values of the optical density of the medium at the boundary of the boundary layer; the results of the calculations are shown in Table 2.

An analysis of the cited graphs leads to the following conclusions:

a. As the optical density $k\delta$ of the gray medium is increased, the radiant energy flux in the boundary layer is diminished.

b. At thermodynamic equilibrium the total heat energy flux at the wall is equal to zero.

c. The total heat flux can be determined from the relations (16) and (17) derived for Q_r and Q_k.

COMPOSITE HEAT TRANSFER AND VISCOUS FRICTION IN A MOVING GRAY MEDIUM WITH LARGE OPTICAL DENSITY

P. K. Konakov, V. T. Kumskov, Yu. P. Sidorov, and V. S. Sidorov

In a number of thermal power plants the working medium has a high optical density, in which case the photon mean free path is commensurate with the molecular mean free path.

The present article discusses the problem of composite heat transfer and viscous friction of a moving gray medium with a large optical density.

Let an incompressible gray medium with a large optical density flow along a plate at velocity w_0 (see the figure). Near the surface of the plate is formed a laminar boundary layer of thickness $\delta = \varphi(x)$.

For nonisothermal motion of the medium at a temperature T_0 a temperature boundary layer of thickness $\delta_T = \varphi(x)$ is also formed near the wall.

The physical characteristics of the medium are its density ρ, coefficient of thermal conductivity λ, absorption coefficient k, and kinematic viscosity coefficient ν, all of which are assumed constant.

The transfer of thermal energy in the gray medium is governed by its convective and radiative components. For this reason the energy equation must be augmented with a term to account for radiative heat transfer.

Making use of the diffusion character of radiant energy transfer in an absorbing and emitting medium, the radiative component of the composite heat transfer is represented as follows:

$$\frac{\sigma_0}{\kappa} \cdot \frac{\partial^2 T^4}{\partial y^2}.$$

The equations for the laminar boundary layer of the gray medium, taking this into account, can be written

$$\frac{\partial w_x}{\partial x} + \frac{\partial w_y}{\partial y} = 0, \tag{1}$$

$$w_x \frac{\partial w_x}{\partial x} + w_y \frac{\partial w_x}{\partial y} = \nu \frac{\partial^2 w_x}{\partial y^2}, \tag{2}$$

$$w_x \frac{\partial T}{\partial x} + w_y \frac{\partial T}{\partial y} = a \frac{\partial^2 T}{\partial y^2} + \frac{\sigma_0}{\kappa} \cdot \frac{\partial^2 T}{\partial y^2}. \tag{3}$$

Bringing in considerations of the similarity of velocity and temperature in the boundary layer, we introduce the following functions:

$$w_x = w_0 \varphi(\eta) \qquad \text{and} \qquad T_x = T_0 \psi(\eta), \tag{4}$$

where

$$\eta = \frac{y}{\delta}, \qquad \eta_T = \frac{y}{\delta_T};$$

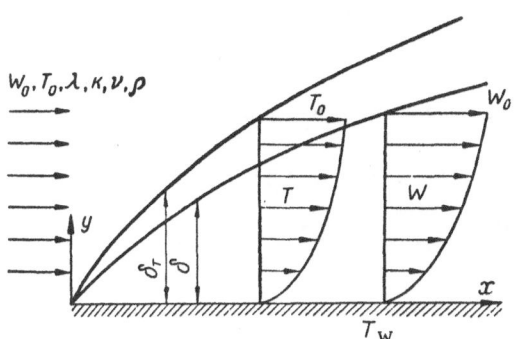

Diagram of the hydrodynamic and temperature boundary layers.

37

δ and δ_T are the thicknesses of the hydrodynamic and temperature boundary layers, respectively.

Equations (4) permit the system of equations (1), (2), and (3) to be written in the form

$$\frac{d\,[\varphi\,(\eta)\,\theta]}{d\eta}\int_0^\eta \varphi\,(\eta)\,d\eta = A_1\frac{d^2\,[\varphi\,(\eta)\,\theta]}{d\eta^2}\,, \tag{5}$$

$$-\frac{d\theta}{d\eta_T}\int_0^{\eta_T}\varphi\,(\eta_T)\,d\eta_T = A_2\frac{d^2\theta}{d\eta_T^2} + A_3\frac{d^2\theta^4}{d\eta_T^2}\,, \tag{6}$$

where

$$A_1 = \frac{\nu}{w_0\delta}\,\frac{1}{\dfrac{d\delta}{dx}}\,, \quad A_2 = \frac{\lambda}{\rho c_p w_0\delta}\,\frac{1}{\dfrac{d\delta}{dx}}\,,$$

$$A_3 = \frac{\sigma_0}{\kappa w_0\delta}\,\frac{1}{\dfrac{d\delta}{dx}}\,, \qquad \theta = \frac{T}{T_0 - T_W}\,.$$

The solution of Eqs. (5) and (6) will be sought in the form of power series, which we put in the following form for the velocity and temperature fields, respectively:

$$\varphi\,(\eta) = a_0 + a_1\eta + a_2\eta^2 + a_3\eta^3 + a_4\eta^4 + \ldots \tag{7}$$

$$\theta = b_0 + b_1\eta_T + b_2\eta_T^2 + b_3\eta_T^3 + b_4\eta_T^4 + \ldots \tag{8}$$

Let us analyze Eq. (5). The relation between the variables η and η_T can be expressed as

$$\eta_T = \eta\xi_T.$$

Multiplying the series (7) and (8), we obtain

$$\varphi\,(\eta)\,\theta = a_0b_0 + (a_0b_1\xi_T + a_1b_0)\,\eta + (a_0b_2\xi_T^2 + a_1b_1\xi_T + a_2b_0)\,\eta^2 +$$
$$+ (a_0b_3\xi_T^3 + a_1b_2\xi_T^2 + a_2b_1\xi_T + a_3b_0)\,\eta^3 + (a_0b_4\xi_T^4 + a_1b_3\xi_T^3 +$$
$$+ a_2b_2\xi_T^2 + a_3b_1\xi_T + a_4b_0)\,\eta^4 + (a_0b_5\xi_T^5 + a_1b_4\xi_T^4 + a_2b_3\xi_T^3 +$$
$$+ a_3b_2\xi_T^2 + a_4b_1\xi_T + a_5b_0)\,\eta^5 + \ldots$$

Differentiating the product $\varphi(\eta)\theta$ with respect to η,

$$\frac{d\,[\varphi\,(\eta)\,\theta]}{d\eta} = (a_0b_1\xi_T + a_1b_0) + 2(a_0b_2\xi_T^2 + a_1b_1\xi_T + a_2b_0)\,\eta +$$
$$+ 3\,(a_0b_3\xi_T^3 + a_1b_2\xi_T^2 + a_2b_1\xi_T + a_3b_0)\,\eta^2 + 4\,(a_0b_4\xi_T^4 + a_1b_3\xi_T^3 +$$
$$+ a_2b_2\xi_T^2 + a_3b_1\xi_T + a_4b_0)\,\eta^3 + 5\,(a_0b_5\xi_T^5 + a_1b_4\xi_T^4 + a_2b_3\xi_T^3 +$$
$$+ a_3b_2\xi_T^2 + a_4b_1\xi_T + a_5b_0)\,\eta^4 + \ldots$$

A second differentiation yields the following results:

$$\frac{d^2\,[\varphi\,(\eta)\cdot\theta]}{d\eta^2} = 2\,(a_0b_3\xi_T^2 + a_1b_1\xi_T + a_2b_0) + 6\,(a_0b_3\xi_T^3 + a_1b_2\xi_T^2 +$$
$$+ a_2b_1\xi_T + a_3b_0)\,\eta + 12\,(a_0b_4\xi_T^4 + a_1b_3\xi_T^3 + a_2b_2\xi_T^2 + a_3b_1\xi_T +$$

$$+ a_4 b_0) \, \eta^2 + 20 \left(a_0 b_5 \xi_\tau^5 + a_1 b_4 \xi_\tau^4 + a_2 b_3 \xi_\tau^3 + a_3 b_2 \xi_\tau^2 + a_4 b_1 \xi_\tau + a_5 b_0 \right) \eta^3 + \ldots$$

It is apparent that the integral of the function $\varphi(\eta)$ can be written

$$\int_0^\eta \varphi(\eta) \, d\eta = a_0 \eta + \frac{a_1}{2} \eta^2 + \frac{a_2}{3} \eta^3 + \frac{a_3}{4} \eta^4 + \frac{a_4}{5} \eta^5 + \ldots$$

With the above expressions in mind, Eq. (5) can be represented as follows:

$$- [c_0 + 2c_1 \eta + 3c_2 \eta^2 + 4c_3 \eta^3 + \ldots] \left(a_0 \eta + \frac{a_1}{2} \eta^2 + \frac{a_2}{3} \eta^3 + \right.$$

$$\left. + \frac{a_3}{4} \eta^4 + \ldots \right) = 2A_1 c_1 + 6A_1 c_2 \eta + 12A_1 c_3 \eta^2 + 20A_1 c_4 \eta^3 + \ldots,$$

where

$$c_0 = a_0 b_1 \xi_\tau + a_1 b_0,$$

$$c_1 = a_0 b_2 \xi_\tau^2 + a_1 b_1 \xi_\tau + a_2 b_0,$$

$$c_2 = a_0 b_3 \xi_\tau^3 + a_1 b_2 \xi_\tau^2 + a_2 b_1 \xi_\tau + a_3 b_0,$$

$$c_3 = a_0 b_4 \xi_\tau^4 + a_1 b_3 \xi_\tau^3 + a_2 b_2 \xi_\tau^2 + a_3 b_1 \xi_\tau + a_4 b_0,$$

$$\cdots \cdots \cdots \cdots \cdots \cdots \cdots \cdots \cdots \cdots$$

or, finally,

$$- \left[c_0 a_0 \eta + \left(2c_1 a_0 + \frac{c_0 a_1}{2} \right) \eta^2 + \left(3c_2 a_0 + c_1 a_1 + \frac{c_0 a_2}{3} \right) \eta^3 + \right.$$

$$+ \left(4c_3 a_0 + \frac{3}{2} c_2 a_1 + \frac{2}{3} c_1 a_2 + \frac{1}{4} c_0 a_3 \right) \eta^4 + \left(5c_4 a_0 + 2c_3 a_1 + \right.$$

$$\left. + c_2 a_2 + \frac{1}{2} c_1 a_3 + \frac{1}{5} c_0 a_4 \right) \eta^5 + \ldots \bigg] = 2A_1 c_1 + 6A_1 c_2 \eta + 12A_1 c_3 \eta^2 +$$

$$+ 20A_1 c_4 \eta^3 + 30A_1 c_5 \eta^4 + \ldots \tag{9}$$

A comparison of the coefficients of identical powers leads to the system of equations

$$\left. \begin{array}{l} c_1 = 0, \\ c_0 a_0 + 6A_1 c_2 = 0, \\ 2c_1 a_0 + \dfrac{1}{2} c_0 a_1 + 12A_1 c_3 = 0, \\ 3c_2 a_0 + c_1 a_1 + \dfrac{1}{3} c_0 a_2 + 20A_1 c_4 = 0, \\ 4c_3 a_0 + \dfrac{3}{2} c_2 a_1 + \dfrac{2}{3} c_1 a_2 + \dfrac{1}{4} c_0 a_3 + 30A_1 c_5 = 0. \end{array} \right\} \tag{10}$$

In order to solve the system (10) and find the coefficients, we formulate the boundary conditions. Inasmuch as the velocity at the wall is equal to zero, from Eqs. (7), with $\eta = 0$, we obtain

$$| \varphi(\eta) |_{\eta=0} = a_0 = 0.$$

Recognizing that the wall temperature is equal to T_W, it follows from Eq. (8) that

$$|\theta|_{\eta=0} = \frac{T_W}{T_0 - T_W} = b_0.$$

Applying the above boundary conditions and the system of equations (10), we obtain the following expressions:

$$
\left.\begin{aligned}
c_0 &= a_1 \frac{T_W}{T_0 - T_W}\,, \\[4pt]
c_1 &= a_1 b_1 \xi_T + a_2 \frac{T_W}{T_0 - T_W} = 0, \\[4pt]
c_2 &= a_1 b_2 \xi_T^2 + a_2 b_1 \xi_T + a_3 \frac{T_W}{T_0 - T_W} = 0, \\[4pt]
\frac{T_W}{T_0 - T_W} &\cdot \frac{a_1^2}{2} + 12 A_1 (a_1 b_3 \xi_T^3 + a_2 b_2 \xi_T^2 + a_3 b_1 \xi_T + a_4 b_0) = 0, \\[4pt]
\frac{T_W}{T_0 - T_W} &\cdot \frac{a_1 a_2}{3} + 20 A_1 (a_1 b_4 \xi_T^4 + a_2 b_3 \xi_T^3 + a_3 b_2 \xi_T^2 + a_4 b_1 \xi_T + a_5 b_0) = 0.
\end{aligned}\right\} \tag{11}
$$

Let us now consider the energy transfer equation (6), representing it also as power series functions. We express the terms of the equation in the following series form:

$$\frac{d\theta}{d\eta_T} = b_1 + 2b_2 \eta_T + 3b_3 \eta_T^2 + 4b_4 \eta_T^3 + 5b_5 \eta_T^4 + \ldots$$

$$\frac{d^2\theta}{a\eta_T^2} = 2b_2 + 6b_3 \eta_T + 12b_4 \eta_T^2 + 20b_5 \eta_T^3 + \ldots$$

$$\int_0^{\eta_T} \varphi(\eta_T)\, d\eta_T = a_0 \eta_T + \frac{a_1}{2} \eta_T^2 + \frac{a_3}{3} \eta_T^3 + \frac{a_3}{4} \eta_T^4 + \ldots$$

$$\frac{d^2\theta^4}{d\eta^2} = 2\left(4b_0^3 b_2 + 6b_0^2 b_1^2\right) + 6\left(4b_0^3 b_3 + 12b_0^2 b_1 b_2 + 4b_0 b_1^3\right)\eta_T +$$

$$+ 12\left(4b_0^3 b_4 + 12 b_0^2 b_1 b_3 + 6b_0^2 b_2^2 + 12b_0 b_1^2 b_2 + b_1^4\right)\eta_T^2 + 20\left(4b_0^3 b_5 + \right.$$

$$\left. + 12b_0^2 b_1 b_4 + 12b_0^2 b_2 b_3 + 12b_0 b_1^2 b_3 + 12b_0 b_1 b_2^2 + 4b_1^3 b_2\right)\eta_T^3 + \ldots$$

These expressions are used to express Eq. (6) in the form

$$
-\left[b_1 a_0 \eta_T + \left(2b_2 a_0 + \frac{b_1 a_1}{2}\right)\eta_T^2 + \left(3b_3 a_0 + a_1 b_2 + \frac{b_1 a_2}{3}\right)\eta_T^3 + \right.
$$

$$
\left. + \left(4b_4 a_0 + \frac{3}{2} b_3 a_1 + \frac{2}{3} b_2 a_2 + \frac{a_3 b_1}{4}\right)\eta_T^4 + \ldots\right] = \tag{12}
$$

$$
= A_2 \left[2b_2 + 6b_0^3 \eta_T + 12b_4 \eta_T^2 + 20b_5 \eta_T^3 + \ldots\right] + A_3 \left[2\left(4b_0^3 b_2 + \right.\right.
$$

$$
\left. + 6b_0^2 b_1^2\right) + 6\left(4b_0 b_3 + 12b_0^2 b_1 b_2 + 4b_0 b_1^3\right)\eta_T + 12\left(4b_0^3 b_4 + 12b_0^2 b_1 b_3 + \right.
$$

$$
\left.\left. + 6b_0^2 b_2^2 + 12b_0 b_1^2 b_2 + b_1^4\right)\eta_T^2 + \ldots\right].
$$

From a comparison of the coefficients of like powers in Eq. (12) we obtain

$$2A_2b_2 + 2A_3\,(4b_0^3b_2 + 6b_0^2b_1^2) = 0, \quad b_1a_0 + 6A_2b_3 + 6A_3\,(4b_0^3b_3 + 12b_0^2b_1b_2 + 4b_0b_1^3) = 0,$$
$$2b_2a_0 + \frac{b_1a_1}{2} + 12A_2b_4 + 12A_3\,(4b_0^3b_4 + 12b_0^2b_1b_3 + 6b_0^2b_2^2 + 12b_0b_1^2b_2 + b_1^4) = 0. \tag{13}$$

Solving the systems of equations of like powers in Eq. (11) and (13) simultaneously, the coefficients of the series (7) and (8) can be written as functions of a_1 and b_1:

$$a_2 = -\frac{a_1b_1\xi_T}{\dfrac{T_W}{T_0 - T_W}}\,,$$

$$a_3 = \frac{a_1b_1^2\xi_T^2}{\left(\dfrac{T_W}{T_0 - T_W}\right)^2} + \frac{6a_1b_1^2A_3\left(\dfrac{T_W}{T_0 - T_W}\right)^2\xi_T^2}{A_2 + 4A_3\left(\dfrac{T_W}{T_0 - T_W}\right)^3}\,,$$

$$a_4 = -\frac{a_1^2}{24A_1} - \frac{72A_3^2a_1b_1^3\left(\dfrac{T_W}{T_0 - T_W}\right)^3\xi_T^3}{\left[A_2 + 4A_3\left(\dfrac{T_W}{T_0 - T_W}\right)^3\right]^2} - \frac{8A_3a_1b_1^3\xi_T^3}{A_2 + 4A_3\left(\dfrac{T_W}{T_0 - T_W}\right)^3} - \frac{a_1b_1^3\xi_T^3}{\left(\dfrac{T}{T_0 - T_W}\right)^3}\,,$$

$$b_2 = -\frac{6A_3b_1^2\left(\dfrac{T_W}{T_0 - T_W}\right)^2}{A_2 + 4A_3\left(\dfrac{T_W}{T_0 - T_W}\right)^3}\,,$$

$$b_3 = -\frac{72A_3^2b_1^3\left(\dfrac{T}{T_0 - T_W}\right)^4}{\left[A_2 + 4A_3\left(\dfrac{T_W}{T_0 - T_W}\right)^3\right]^2} - \frac{4A_3b_1^3\left(\dfrac{T_W}{T_0 - T_W}\right)}{A_2 + 4A_3\left(\dfrac{T_W}{T_0 - T_W}\right)^3}\,,$$

$$b_4 = -\frac{a_1b_1}{24\left[A_2 + 4A_3\left(\dfrac{T_W}{T_0 - T}\right)^3\right]^2} - \frac{1080A_3^3b_1^4\left(\dfrac{T_W}{T_0 - T_W}\right)^6}{\left[A_2 + 4A_3\left(\dfrac{T_W}{T_0 - T_W}\right)^3\right]^3} +$$
$$+ \frac{120A_3^2b_1^4\left(\dfrac{T_W}{T_0 - T_W}\right)^3}{\left[A_2 + 4A_3\left(\dfrac{T_W}{T_0 - T_W}\right)^3\right]^2} - \frac{A_3b_1^4}{A_2 + 4A_3\left(\dfrac{T_W}{T_0 - T_W}\right)^3}\,.$$

Analyzing the coefficients of different powers of the variable η, we obtain the following expressions for the variation in velocity and temperature in the boundary layer:

$$\frac{\rho w_x}{\rho_0 w_0} = a_1\eta - \frac{a_1b_1\xi_T}{\dfrac{T_W}{T_0 - T_W}}\,\eta^2 + \frac{a_1b_1^2\xi_T^2}{\left(\dfrac{T_W}{T_0 - T_W}\right)^2}\,\eta^3 +$$

$$+ \frac{6A_3a_1b_1^2\xi_T^2\left(\dfrac{T_W}{T_0 - T_W}\right)}{A_2 + 4A_3\left(\dfrac{T_W}{T_0 - T_W}\right)^3}\,\eta^3 - \frac{a_1^2}{24A_1}\,\eta^4 - \frac{a_1b_1^3\xi_T^3}{\left(\dfrac{T_W}{T_0 - T_W}\right)^3}\,\eta^4 + \cdots \tag{14}$$

$$\frac{T}{T_0 - T_W} = \frac{T_W}{T_0 - T_W} + b_1 \xi_T \eta - \frac{a_1 b_1 \xi_T^4}{24 \left[A_2 + 4A_3 \left(\frac{T_W}{T_0 - T_W} \right)^3 \right]} \eta^4 + \cdots$$ (15)

From the boundary conditions for the temperature boundary layer with $\eta = 1$ we obtain

$$\left| \frac{T - T_W}{T_0 - T_W} \right|_{\eta=1} = 1 \quad \text{and} \quad \left| \frac{dT}{d\eta} \right|_{\eta=1} = 0.$$ (16)

With due regard for Eq. (16), Eq. (15) can be represented by the relations

$$\left.\begin{aligned} b_1 \xi_T - \frac{b_1 a_1 \xi_T^4}{24 \left[A_2 + 4A_3 \left(\frac{T_W}{T_0 - T_W} \right)^3 \right]} &= 1 \\[2em] b_1 \xi_T - \frac{b_1 a_1 \xi_T^4}{6 \left[A_2 + 4A_3 \left(\frac{T_W}{T_0 - T_W} \right)^3 \right]} &= 0. \end{aligned}\right\}$$ (17)

The solution of the system (17), in the form

$$b_1 \xi_T = \frac{4}{3}$$

is substituted into Eq. (14), whereupon

$$\frac{\rho w_x}{\rho_0 w_0} = a_1 \eta - a_1 \vartheta_0 \eta^2 + a_1 \vartheta_0^2 \eta^3 + \frac{\frac{32}{3} a_1 \frac{T}{T_0 - T_W}}{\chi + 4 \left(\frac{T_W}{T_0 - T_W} \right)^3} \eta^3 - \frac{a_1^2}{24 A_1} \eta^4 - a_1 \vartheta_0^3 \eta^4,$$ (18)

where

$$\vartheta_0 = \frac{4}{3} \cdot \frac{T_0 - T_W}{T_W} \quad \text{and} \quad \chi = \frac{A_2}{A_3} = \frac{\lambda \kappa}{\rho c_p \sigma_0}.$$

From the boundary conditions for the hydrodynamic boundary layer with $\eta = 1$ we obtain

$$\left| \frac{\rho w_x}{\rho_0 w_0} \right|_{\eta=1} = 1, \quad \left| \frac{d(\rho w_x)}{d\eta} \right|_{\eta=1} = 0.$$ (19)

With these expressions in mind, Eq. (14) is represented by the following relations:

$$\left.\begin{aligned} a_1 - a_1 \vartheta_0 + a_1 \vartheta_0^2 - a_1 \vartheta_0^3 + \frac{\frac{32}{3} a_1 \frac{T_W}{T_0 - T_W}}{\chi + 4 \left(\frac{T_W}{T_0 - T_W} \right)^3} - \frac{a_1^2}{24 A_1} &= 1, \\[2em] a_1 - 2a_1 \vartheta_0 + 3a_1 \vartheta_0^2 - 4a_1 \vartheta_0^3 + \frac{32 a_1 \frac{T_W}{T_0 - T_W}}{\chi + 4 \left(\frac{T_W}{T_0 - T_W} \right)^3} - \frac{a_1^2}{24 A_1} &= 0. \end{aligned}\right\}$$ (20)

Solving the system of equations (20) enables us to obtain the values of the coefficients a_1 and A_1:

$$a_1 = \frac{4}{3 - 2\vartheta_0 + \vartheta_0^2 + \frac{\frac{32}{3} \cdot \frac{T_W}{T_0 - T_W}}{\chi + 4 \left(\frac{T_W}{T_0 - T_W} \right)^3}},$$ (21)

$$A_1 = \frac{a_1^2}{24\left(3 - 2a_1 + a_1\vartheta_0 - a_1\vartheta_0^3\right)} \; ; \tag{22}$$

then, since it has been assumed beforehand that $A_1 = \dfrac{\mu}{\rho w_0 \delta \dfrac{d\delta}{dx}}$, this expression can be compared with the

value obtained for the coefficient (22) to find a formula for the thickness of the hydrodynamic boundary layer:

$$\delta = \sqrt{\frac{2\mu x}{\rho w_0 A_1}} = \sqrt{\frac{48\mu x \left(3 - 2a_1 + a_1\vartheta_0 - a_1\vartheta_0^3\right)}{\rho w_0 a_1^2}} \, , \tag{23}$$

and the expressions (14) and (15) for the velocity and temperature fields are finally written in the form

$$\frac{\rho w_x}{\rho_0 w_0} = \frac{a_1}{\xi}\,\eta - \frac{a_1\vartheta_0}{\xi^2}\,\eta^2 + \frac{a_1\vartheta_0^2}{\xi^3}\,\eta^3 + \frac{\dfrac{32}{3}a_1\dfrac{T_W}{T_0 - T_W}}{x + 4\left(\dfrac{T_W}{T_0 - T_W}\right)^3}\,\eta^3 - \frac{a_1\vartheta_0^3}{\xi^4}\,\eta^4 - \frac{a_1^2}{24A_1\xi^4}\,\eta^4 + \ldots, \tag{24}$$

$$\frac{T - T_W}{T_0 - T_W} = \frac{\dfrac{4}{3}}{\xi_T}\,\eta_T - \frac{a_1}{18\left[A_2 + 4A_3\left(\dfrac{T_W}{T_0 - T_W}\right)^3\right]\xi_T^4}\,\eta_T^4 + \ldots \tag{25}$$

Making use of Eqs. (24) and (25), we obtain analytical relations for the viscous friction and composite heat transfer of a gray medium with a large optical density moving past a plate.

To obtain an analytical relation for the viscous friction, we make use of the Newtonian hypothesis

$$S_W = \mu \frac{dw}{dy} \, , \tag{26}$$

where μ is the dynamic viscosity.

On the basis of the relation (24), recalling that $\rho = P/RT$, and stopping with the first two terms of the series, we obtain the following expression for the velocity along the x axis:

$$w_x = \frac{a_1}{\xi} \cdot \frac{y}{\delta}\,\rho_0 w_0 \frac{RT}{P} - \frac{a_1^2}{24A_1\xi^4}\left(\frac{y}{\delta}\right)^4 \rho_0 w_0 \frac{RT}{P} \, . \tag{27}$$

With Eq. (25) we can write the dependence (27) in the form

$$w_x = \rho_0 w_0 \frac{R}{P} \cdot \frac{a_1}{\xi} T_W \frac{y}{\delta} + \frac{4}{3}\rho_0 w_0 \frac{R}{P} \cdot \frac{a_1}{\xi^2}(T_0 - T_W)\left(\frac{y}{\delta}\right)^2 + \cdot$$
$$+ \frac{\rho_0 w_0 T_W a_1^2 R}{24A_1\xi^4 P}\left(\frac{y}{\delta}\right)^4 - \frac{4\rho_0 w_0 R (T_0 - T_W) a_1^2}{72A_1\xi^5 P}\left(\frac{y}{\delta}\right)^5 \, . \tag{28}$$

Substituting the value (28) into Eq. (26), we obtain an equation for evaluating the viscous friction:

$$S_W = \mu\left[\rho_0 w_0 \frac{Ra_1 T_W}{P\xi\delta} + \frac{8}{3}\rho_0 w_0 \frac{Ra_1 (T_0 - T_W)}{P\xi^2\delta^2}\,y - \frac{1}{6}\rho_0 w_0 \frac{Ra_1^2 T_W}{PA_1\xi^4\delta^4}\,y^3 + \ldots\right]. \tag{29}$$

We next obtain analytical expressions for the heat fluxes at the wall in composite heat transfer. It is known that the radiative component of composite heat transfer can be expressed as

$$q_{\mathrm{r}} = \frac{\sigma_0}{\dfrac{1}{A_{\mathrm{W}}} - \dfrac{1}{2}}\left(T_\delta^4 - T_{\mathrm{W}}^4\right), \tag{30}$$

where T_δ is the temperature of the equilibrium layer.

The temperature of the equilibrium can be determined from Eq. (25), assuming $\eta = l_{\mathrm{s}}/\delta = 1/k\delta$,

$$T_0 = T_{\mathrm{W}} + \frac{4}{3}\cdot\frac{(T_0 - T_{\mathrm{W}})}{\xi\xi_{\mathrm{T}}\varkappa\delta} - \frac{a_1\,(T_0 - T_{\mathrm{W}})}{18\left[A_2 + 4A_3\left(\dfrac{T_{\mathrm{W}}}{T_0 - T_{\mathrm{W}}}\right)^3\right]\xi_{\mathrm{T}}\xi^4\,(\varkappa\delta)^4} + \cdots \tag{31}$$

It is assumed in this case that the photon mean free path l_{s} is commensurate with the molecular mean free path l_{m}. Taking Eq. (31) into account, Eq. (30) can be represented in the form

$$q_{\mathrm{r}} = \frac{\delta_0}{\dfrac{1}{A_{\mathrm{W}}} - \dfrac{1}{2}}\left[\frac{16T_{\mathrm{W}}^3\,(T_0 - T_{\mathrm{W}})}{3\xi\xi_{\mathrm{T}}\varkappa\delta} + \frac{32T_{\mathrm{W}}^2\,(T_0 - T_{\mathrm{W}})^2}{3\xi^2\xi_{\mathrm{T}}^2\,(\varkappa\delta)^2} + \right.$$
$$\left. + \frac{256T_{\mathrm{W}}\,(T_0 - T_{\mathrm{W}})^3}{27\xi^3\xi_{\mathrm{T}}^3\,(\varkappa\delta)^3} + \frac{256\,(T_0 - T_{\mathrm{W}})^4}{81\xi^4\xi_{\mathrm{T}}^4\,(\varkappa\delta)^4} + \cdots\right]. \tag{32}$$

For the convective component of composite heat transfer at the wall we write

$$q_{\varkappa} = -\lambda\left|\frac{dT}{dy}\right|_{\mathrm{W}}, \tag{33}$$

and, taking Eq. (25) into account for $\eta = 0$, we obtain

$$q_{\varkappa} = \frac{4\lambda\,(T_0 - T_{\mathrm{W}})}{3\xi\xi_{\mathrm{T}}\delta}. \tag{34}$$

Equations (33) and (34) can be used to determine the total heat flux in composite heat transfer.

We write the ratio of the radiative component to the convective component in composite heat transfer in a gray medium:

$$\frac{q_{\mathrm{r}}}{q_{\varkappa}} = \frac{\sigma_0}{\varkappa\lambda\left(\dfrac{1}{A_{\mathrm{W}}} - \dfrac{1}{2}\right)}\left[4T_{\mathrm{W}}^3 + \frac{8T_{\mathrm{W}}^2\,(T_0 - T_{\mathrm{W}})}{\xi\xi_{\mathrm{T}}\varkappa\delta} + \frac{64T_{\mathrm{W}}\,(T_0 - T_{\mathrm{W}})^2}{9\xi^2\xi_{\mathrm{T}}^2\,(\varkappa\delta)^2} + \frac{64\,(T_0 - T_{\mathrm{W}})^3}{27\xi^3\xi_{\mathrm{T}}^3\,(\varkappa\delta)^3} + \cdots\right].$$

For a large optical density $k\delta$ Eq. (35) is simplified to

$$\frac{q_{\mathrm{r}}}{q_{\varkappa}} = \frac{4\sigma_0 T_{\mathrm{W}}^3}{\varkappa\lambda\left(\dfrac{1}{A_{\mathrm{W}}} - \dfrac{1}{2}\right)}.$$

Conclusions

1. We have made use of the laminar boundary layer equations, augmenting the energy conservation equation with a term to account for the radiative component of composite transfer. Analytical relations have been

derived on the basis of these equations, using power series, for the viscous friction and heat transfer in the motion of gray media.

2. The variation of the optical density $k\delta$ of the medium considerably affects the value of the radiative component of composite heat transfer.

3. An analysis of the solutions obtained herein shows that the hydrodynamic state is the decisive factor in the intensification, of not only the convective but also the total composite heat transfer. A variation in the radiative component is manifested only slightly in the hydrodynamics of the process.

4. For gray media with a very high optical density ($k\delta$ large) the fraction of radiant energy transfer is lessened.

Literature Cited

1. P. K. Konakov, S. S. Filimonov, and B. A. Khrustalev, Heat Transfer in Boiler Combustion Chambers (Rechizdat, 1960).

THE ABSORPTION COEFFICIENT OF A GRAY MEDIUM

V. T. Kumskov and Yu. P. Sidorov

The investigation of problems involving composite heat transfer in moving gray media is related to the investigation of the emissive properties of these media. The emissive properties include the absorptive and emissive powers of the medium, as expressed in terms of the absorption coefficient k and self-radiation coefficient (emittance) η.

The absorptive power of a medium was first interpreted by Bouguer in his treatise on the gradation of light [1]: "The forces possessed by light after transmission through different thicknesses may be represented by the ordinates of a logarithmic curve whose axis is the thickness of the body. The logarithmic curve will differ, depending on the transparency of the body."

The geometric interpretation of Bouguer's hypothesis, according to which the variation in radiation intensity is equal to $I = I_0 e^{-kl}$, is illustrated in Fig. 1.

According to Bouguer's hypothesis, the transmittivity of a medium is defined as a quantity proportional to $1/k$.

In most works the absorption coefficient k is also interpreted as a coefficient of proportionality, taking into account the attenuation of radiant energy transmitted through an absorbing medium. We will examine the mechanism of radiant energy transfer in an absorbing medium with constant absorption coefficient k.

Let us suppose that the intensity after transmission through an element of the medium of thickness l changes from I_1 to I_2. In accordance with Bouguer's hypothesis, for a layered element ds, we can write

$$\frac{dI}{ds} = \kappa I. \tag{1}$$

The solution of this equation within the limits of the layered element l yields the result

$$\frac{I_2}{I_1} = e^{-kl} \quad \text{or} \quad \frac{I_1 - I_2}{I_1} = e^{-kl}. \tag{2}$$

The left-hand side of Eq. (2) represents the ratio of absorbed energy to the initial intensity of the radiant energy and expresses the absorptive power of the medium, its emissivity:

$$\varepsilon = 1 - e^{-kl}.$$

The emissivity is determined experimentally as a function of the total pressure p, concentration of the gas mixture c, temperature T, and thickness of the gas layer l. The resultant experimental data for the emissivity of the layer permit the absorption coefficient $k = \varphi(p, c, T, l)$ to be determined [2].

The experimental data of Hottel [3] and the relation $\varepsilon_g = 1 - e^{-kl}$ for the absorptive power of CO_2 and H_2O can be used to numerically determine the value of the absorption coefficient k. For example, in the case of a combustion chamber:

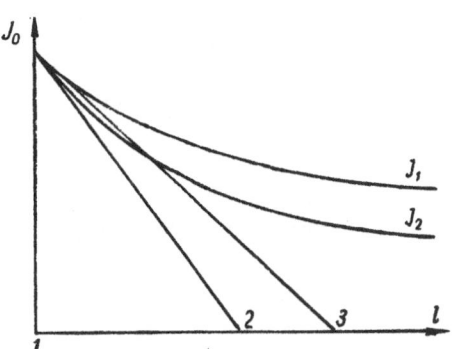

Fig. 1. Geometric interpretation of Bouguer's law.

$l_s = 0.552$ m, $T = 1500°C$, $\rho_g = 1.76$ kg/Nm^3, and $pl = 0.975$; for $\varepsilon_{CO_2} = 0.25$, the absorption coefficient $k_{CO_2} = 0.292$; for $\varepsilon_{H_2O} = 0.52$, the absorption coefficient $k_{H_2O} = 1.51$.

The photon mean free path can be determined from the total absorption coefficient $k = 2$.

$$l_s = \frac{1}{2}\ \text{m.}$$

The presence of a large quantity of fine suspended dust particles (soot, ash) in the gas makes the medium more optically dense. The given medium may be regarded as a gray body, since it is invested with the capability of absorbing radiant energy over the entire spectral band.

Let us investigate a dusty medium with a particle mass concentration c. We denote the density of the medium as ρ and the particle diameter as d. A layer of the dusty medium presents a shield against the transmission of radiant energy. The number of particles in a layered element ds (assuming the other two dimensions of the element to be unity) can be determined according to the expression

$$N = \frac{cds}{G} = \frac{cds}{\frac{\pi}{6}d^3\rho}. \tag{3}$$

The intensity of radiant energy passing through the dusty medium is attenuated from I_1 to I_2. The change in intensity in this case is determined by the shielding effect. The total shielding surface of the particles is determined by the product of their number and mean cross section:

$$Nf = \frac{cds}{\frac{\pi}{6}d^3\rho} \cdot \frac{\pi d^2}{4},$$

hence

$$dI = -IN\frac{\pi d^2}{4}. \tag{4}$$

On the other hand, the change in radiation intensity is expressed by Eq. (1). The solution of Eqs. (1) and (4) makes it possible to determine the attenuation factor of the dusty gaseous medium:

$$\kappa_d = \frac{3}{2} \cdot \frac{c}{d\rho}. \tag{5}$$

As we see, the attenuation factor increases with increasing mass concentration c and with decreasing particle diameter d and density ρ.

The total absorption coefficient k in the dust-filled currents of combustion products depends on the coefficients of absorption of radiant energy by triatomic gases, k_g, and by solid dust particles, k_d:

$$\kappa = \kappa_g + \kappa_d.$$

The numerical value of the absorption coefficient k_d can be determined by means of Eq. (5). For example, if the combustion products are impregnated with chrome magnesium dust with a mass concentration $c = G_d/V = 0.227$ kg/m^3 and density $\rho = 2750$ to 2850 kg/m^3, for a mean dust particle diameter $d = 30\ \mu$ we obtain an absorption coefficient $k_d = 6$.

Consequently, the total absorption coefficient of a gas medium impregnated with chrome–magnesium dust particles will be $k = 8$, corresponding to a photon mean free path $l = 0.12$ m.

The above example shows that dusting the combustion products effects an increase in optical density of the medium.

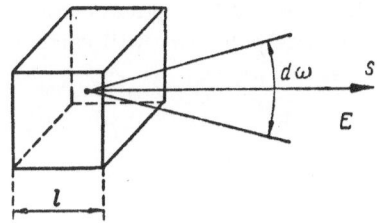

Fig. 2. Diagram of the transfer of radiant energy.

However, a consideration of the radiant energy transport mechanism in media on the basis of Bouguer's hypothesis has a significant drawback, namely that it disregards the exchange of energy between photons and molecules, which brings in an additional self-radiant or emittance effect. In studying the emissive properties of a medium, therefore, it is a good idea to make use of the radiant energy transport equation in a gray medium. Then the change in radiation intensity is an event involving the onset of absorption and emission processes.

Consider a gray medium in thermodynamic equilibrium at a temperature T.

Every elementary cube (Fig. 2) with geometric dimension l radiates the following amount of energy per unit time in the direction s within the solid angle $d\omega$:

$$d^5E = \eta l^3 d\omega. \tag{6}$$

The amount of energy transmitted through the lateral face of the cube in the same period of time and in the same solid angle $d\omega$ is

$$d^4E = I_s l^2 d\omega. \tag{7}$$

The radiation intensity I_s and emittance η are related by the formula

$$I_s = \eta l. \tag{8}$$

An analysis of Eq. (8) leads to the conclusion that at constant emittance η the radiation intensity increases with increasing l. However, this contradicts physical reality, since the quantity I_s is finite.

The radiation intensity increases up to some definite value, and then its growth tends to be offset by absorption. Assuming that the absorption of photons is only possible by material bodies, we can determine the maximum radiation intensity at a given temperature:

$$I_{max} = I_0 = \eta l_s, \tag{9}$$

where l_s is the photon mean free path.

Recognizing that the photon mean free path is inversely proportional to the absorption coefficient coefficient k, Eq. (9) may be written as follows, according to the Kirchhoff law for the state of thermodynamic equilibrium:

$$\frac{\eta}{\kappa} = I_0,$$

where k is the radiant energy attenuation factor (extinction coefficient), which includes both absorption and scattering.

The effect of diffuse scattering in the type of media that are necessarily involved in the calculation of composite heat transfer in thermal power plants is very slight, so that the attenuation effect is determined primarily by diffuse absorption.

For media in other than the state of thermodynamic equilibrium the change in intensity of radiation in its direction of motion will be determined by the difference between the absorbed energy and self-radiation energy:

$$dI_s = -\kappa I_s ds + \eta ds. \tag{10}$$

In analyzing Eq. (10), we note one very important fact. The photon mean free path l_s must be greater than the elemental length differential ds, because otherwise the amount of energy absorbed over the distance ds will be greater than energy radiated by the medium over this same interval.

With these remarks in mind, one can indicate the area of application of Eq. (10). As the optical density kl of the medium is increased the photon mean free path becomes shorter and ultimately the time will come when l_s is smaller than ds. In this case Eq. (10) loses its physical sense.

For a medium in the state of thermodynamic equilibrium we can write

$$\frac{\eta}{\kappa} = \frac{\sigma_0 T^4}{\pi} \quad \text{or} \quad \eta = \kappa I_0,$$

where I_0 is the black body radiation intensity. Then the radiant energy transport equation is written in the form

$$\frac{dI}{ds} = -\kappa I + \kappa I_0.$$

The solution of this equation within the limits of variation of the intensity as it goes from I_1 to I_2 leads to the expression

$$\frac{I_2 - I_0}{I_1 - I_0} = e^{-\kappa l}, \tag{11}$$

Utilizing the experimental data of Hottel, we obtain a value of $k \approx 2$ for the absorption coefficient, which is slightly higher than the value obtained on the basis of Bouguer's law.

Consequently, as a result of determining the absorption coefficient by either of the means indicated, the numerical order of magnitude of k is the same, increasing only slightly with the injection of dust into the flow. The rather long mean free path of the photon indicates that the nonequilibrium state of the furnace medium is characterized by a low optical density.

To determine the radiant flux in media with a large optical density, i.e., media almost in thermodynamic equilibrium, it is essential to make use of gradient concepts. The application of gradient concepts is justified by the fact that the photon mean free path tends in the limit to the molecular mean free path.

Solutions to the problem of radiative heat transfer in emitting-absorbing media with a large optical density are obtained in [4] on the basis of gradient concepts.

In treating the composite heat transfer in glass containers the absorption coefficient k was assumed to be very high, so that the thermal radiation at the bounding surfaces could be considered as black body radiation.

Let us review the gradient approach to evaluating the coefficient k.

In this case the fundamental analytical equation of composite heat transfer for an equilibrium layer may be written in the form

$$q = \frac{\sigma_0 (T_\delta^4 - T_W^4)}{\dfrac{1}{A_W} - \dfrac{1}{2}} + \lambda \kappa (T_\delta - T_W). \tag{12}$$

Utilizing the experimental data on composite heat transfer and determining the convective component, we can find the temperature T. The value of the absorption coefficient k can be calculated at a certain temperature from Eq. (11). The calculations show that the value of k is of the order 10^3, i.e., the photon mean free path l_s is small.

Recently P. K. Konakov, in treating the more general equations of heat and mass transfer in a gray medium [5], obtained the following analytical expression for the absorption coefficient:

$$\kappa = \frac{\rho^2 c^2 c_r T_r}{16 \lambda \sigma_0 T^3}, \tag{13}$$

where c and c_r are the specific heats at T and T_r.

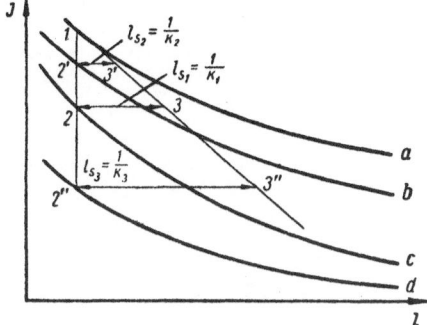

Fig. 3. Variation in radiation intensity I as a function of l.

For the case of radiative equilibrium $T = T_r$, the formula is simplified, assuming the form

$$\kappa = \frac{p^2 c^3}{16 \lambda \sigma_0 T^2} .$$ (14)

In Eq. (13) the coefficient k stands as a measure of the departure of the medium from the state of radiative equilibrium. The same gray medium can have very different numerical values of the absorption coefficient, depending on its degree of nonequilibrium ($T_r \neq T$). As $T_r \to 0$ the absorption coefficient $k \to 0$. For $T_r = T$ the absorption coefficient is very large.

Let us examine the above conclusion in closer detail. To do so, we write Eq. (10) in the form

$$I_s - \frac{\eta}{\kappa} = -\frac{1}{\kappa} \frac{dI_s}{ds} .$$ (15)

The relation (15) is illustrated graphically in Fig. 3; curve a shows the variation in radiation intensity I_s, curves b, c, d show the variation in self-radiation intensity η / k in nonequilibrium states ($T_r \neq T$). The ordinate 1-2 determines the degree of radiative nonequilibrium. In this case the segment 2-3 is expressed geometrically by the quantity

$$l_{s_1} = \frac{1}{\kappa_1} .$$

As the degree of radiative nonequilibrium decreases (curve b for the self-radiation, ordinate 1-2') the segment 2'-3' is geometrically expressed by the quantity $l_{s_2} = 1/k_2$. It is clear from the figure that in this case l_{s_2} decreases and the absorption coefficient k_2 increases, reaching its maxima in the state of radiative equilibrium.

An analysis of Eq. (14) shows that the absorption coefficient k and coefficient of thermal conductivity λ are interdependent quantities.

It is apparent, in determining the absorption coefficient k, that its value varies between very broad limits.

At thermodynamic equilibrium the absorption coefficient is very large, tending to zero as the departure from equilibrium is increased.

It follows from our analysis of the various approaches to estimating the absorption coefficient k of a medium that the absorption coefficient of real media is a very important and complicated quantity; the investigation of its physical essence is a problem of prime significance in the study of composite heat transfer in gray media.

Literature Cited

1. Pierre Bouguer, Optical Treatise on the Gradation of Light [Russian translation] (Izd. Akad. Nauk SSSR, 1950).
2. L. A. Goryainov and V. T. Kumskov, Calculation of the Radiative Component of Composite Heat Transfer, Tr. MIITa (Transactions of the Moscow Railroad Engineers Institute), No. 112 (1959).
3. H. C. Hottel and R. B. Egbert, Trans. Am. Inst. Chem. Engrs., 38:531 (1942).
4. M. Czerny and L. Genzel, Glastech. Ber., 25(8):242; (12):387 (1952).
5. P. K. Konakov, Energy Transport in a Gray Medium, Inzh.-Fiz. Zh., 6(6):(1963).

METHODS FOR THE SEPARATION OF COMPOSITE HEAT TRANSFER

L. A. Goryainov

Heat transfer in a moving emitting medium by convection and radiation is a frequent occurrence in industrial heat exchanging equipment. In this form the effect is usually called composite heat transfer.

It has been shown in a number of recent papers that convective and radiative heat transfer in the composite process are mutually related. This makes it necessary to investigate in detail the nature of the interrelationship in order to learn more about the laws governing the process and to find analytical functions that reflect the physical substance of the effect.

In the investigation of composite heat transfer it is separated into its radiative and convective components.

In the present article we discuss the existing methods for separating composite heat transfer, giving a critical evaluation of the methods and setting forth a number of problem areas in which further research is needed.

Before undertaking our investigation of the methods for separating composite heat transfer into its components, we will consider some of the laws of this process.

The analysis of composite heat transfer on the basis of similarity theory provides a basis for obtaining the following invariant function:

$$\mathrm{Nu}_{tot} = f_1\left(\mathrm{Re}, \mathrm{Pr}, \mathrm{Bu}, \Pi_f, A_w \; \frac{T_w}{T_0}, \chi\right), \tag{1}$$

where $\mathrm{Nu}_{tot} = \alpha_{tot} d/\lambda$ is the composite heat transfer criterion; α_{tot}, α_k, and α_r (where $\alpha_{tot} = \alpha_k + \alpha_r$) are the heat transfer coefficients (total, convective, and radiative), $\mathrm{W/m^2 \cdot deg}$; λ is the coefficient of thermal conductivity, $\mathrm{W/m^2 \cdot deg}$; Re, Pr, Bu, Π_f are the Reynolds, Prandtl, and Bouguer numbers and furnace criterion, A_w is the emissivity of the heat-sensitive surface, χ is a simplex characterizing the geometry of the heat exchanger, T_w/T_0 is the temperature simplex.

It can be shown that the convective component of composite heat transfer is also in general a function of the criteria and simplexes on the right-hand side of the expression (1).

Let us consider in further detail the existing methods for separating composite heat transfer into its radiative and convective components.

1. Air Blast Method

In the application of this method hot air is blasted through the test section and the necessary measurements are made in order to obtain the invariant convective heat transfer relation. As we know, this relation has the general form

$$\mathrm{Nu} = f_2\left(\mathrm{Re}, \mathrm{Pr}, \frac{T}{T_0}, \chi\right), \tag{2}$$

where Nu is the Nusselt number.

The relation is used to determine the convective component of the composite heat transfer and, as the difference between the total and convective flows, the radiative component. It is apparent from Eqs. (1) and

(2) that the air blast method assumes that the criterion Nu is independent of the quantities Bu, Π_f, and A_w for the case of composite heat transfer. Then it is assumed that radiation does not affect the convective component of heat transfer. The flow hydrodynamics governing convective heat transfer can have an effect on the radiative component as well. This is specifically confirmed by the nearly constant ratio of the radiative and convective components obtained in [2, 7, 9].

When the air blast method is used for a specific heat exchanger the air temperature and Prandtl number vary only slightly, while the geometric characteristic does not change at all. Consequently, as a rule, the relation for convective heat transfer is sought in the form

$$Nu = c_1 Re^n Pr^m \tag{3}$$

or

$$Nu = c_2 Re^n. \tag{4}$$

The problem of arriving at a defining temperature has been resolved differently by different researchers. In [1–5] the wall temperature is taken as the defining temperature, in [6] it is the flow temperature, in [7] the boundary layer temperature, and in [8] the temperature at the inlet to the experimental section.

After processing the results of a hot air blast through a test section according to power laws of the type $Nu = CRe^n Pr^m$, with the Reynolds number defined in terms of the linear velocity and mass flow for three values of the defining temperature, we obtain five variant relations (see pp. 91ff). If the Reynolds number is expressed only in terms of the linear velocity, there will be three such relations. If the resultant relations are extended to the case of composite heat transfer, the convective component calculated on the basis of these relations will have a different numerical value. The maximum value of the convective component is obtained using the relation in which the physical parameters are normalized to the mean flow temperature, the minimum value to the wall temperature. With the temperature simplex introduced into the invariant relation, the problem is further complicated. It must be noted that recommendations as to the effect of the temperature simplex in cooling of the medium exist only in the interval $0.5 < T_w/T_0 < 1$, whereas in the investigations considered above this quantity was considerably less than 0.5. It is decided in [10] that there is an advantage to processing the test data for the case of cooling of the medium according to relations in which the physical parameters are normalized to the mean flow temperature. In the actual tests, however, preference is given to the temperature of the wall or boundary layer, and the temperature simplex is left out altogether.

Consequently, the problem of processing the results of air blasting and the choice of a type of analytical relation must be investigated in further detail. It may be stated for now that the air blast method will give a smaller error the lower the optical density of the medium. The radiative component of heat transfer for the case of combustion products flowing in a cooled channel 100 mm in diameter [18], as determined by the air blast method, roughly agrees with the results of a determination based on the notions of a boundary layer in an emitting medium.

With the continued application of air blasting it should be of interest to conduct additional measurements of the viscous friction and to process the results according to relations obtained on the basis of the Reynolds analogy of heat and mass transfer.

2. Method of Measuring the Molecular Temperature of the Medium Near a Heat-Sensitive Surface

Data from the detailed temperature measurements of the medium near a heat-sensitive surface for the case of composite heat transfer can be used to construct a temperature profile and determine the temperature gradient at the wall. In this case the convective component can be calculated from the relation

$$q_\kappa = -\lambda \left| \frac{\partial t}{\partial n} \right|_w, \tag{5}$$

where λ is the thermal conductivity of the medium, $W/m \cdot deg$; $\partial t/\partial n|_w$ is the temperature gradient at the wall, in deg/m.

3. Method of Two Radiometers

This method is based on a measurement of the total heat flux of two instruments identical in form with different surface emissivities. The instruments are placed in succession at the same site under identical conditions. It is then assumed that in both cases the convective component q_k of the heat transfer and incident radiation on the radiometer E_{in} are the same.

The heat fluxes for the radiometers can be determined from the expressions

$$q_1 = A_1 E_{in} - A_1 \sigma_0 T_{w_1}^4 + q_{k'} \tag{6}$$

$$q_2 = A_2 E_{in} - A_2 \sigma_0 T_{w_2}^4 + q_k, \tag{7}$$

where q_1 and q_2 are the specific heat fluxes of the first and second radiometer, W/m^2; A_1 and A_2 are the surface emissivities of the radiometers; T_{w_1} and T_{w_2} are the wall temperatures of the radiometers; σ_0 is the black body radiation constant.

The quantities q_1, q_2, T_{w_1}, T_{w_2}, A_1, and A_2 are measured. Simultaneous solution of Eqs. (6) and (7) enables us to determine the convective component q_k and, as the difference between the total heat flux and convective component, the radiative component q_r.

The quantities A_1 and A_2 are different from each other and from the surface emissivity A_w of the test section near the radiometer placement site. Consequently, with identical incident radiation the reflected radiative fluxes will be different (for different values of A_1, A_2, and A_w).

It is assumed in the application of this method that the variation in flux density of the reflected radiation (radiant energy density at the heat-sensitive surface) does not alter the values of the incident radiation and molecular temperature field at the heat-sensitive surface. The reverse influence of convection on radiation cannot occur. In the event three radiometers are used, Eqs. (6) and (7) must be augmented by the new equation

$$q_3 = A_3 E_{in} - A_3 \sigma_0 T_{w_3}^4 + q_k. \tag{8}$$

The three equations (6), (7), and (8) contain two unknowns E_{in} and q_k. Solving these equations in pairs, (6)–(7), (6)–(8), and (7)–(8), we obtain three values each for E_{in} and q_k. Agreement of the values obtained will confirm the correctness of the assumptions on which the method is based.

The two-radiometer method was used in [11–14, 6, 7, 8], two and three radiometers were used in [15, 16]. In the latter instance, however, the tests disclosed a considerable scatter in experimental points, tending to undermine the conclusion that the two- and three-radiometer methods yield identical results.

The two-radiometer method is a local one, and in order to extend the results obtained to the test section as a whole it is necessary to take an average, which may lead to substantial error.

One of the shortcomings of the method lies in the painstaking care that must be exercised in placement of the instruments, since the slightest placement error in the two successive measurements can produce a sizable resultant error. The value of q_k (or E_{in}) will be adjusted according to the results of measurements in the test section, and the radiative component at the surface emissivity of the test section can be calculated from the values of E_{in} (providing T_w and A_w are known) according to the relation

$$q_r = E_{in.} A_w - \sigma_0 T_w^4 A_w. \tag{9}$$

4. Method of Direct Measurement of the Radiation Output

In the direct measurement of the radiation output an instrument for measuring the incident radiation is placed in a special window of the test section. The circuitry and description of the instruments are given in [7, 17]. The instrument must be precalibrated against a black body and the calibration results used to ascertain the constant of the instrument.

The mean intensity of the incident radiation in the viewing angle of the instrument is found from the mean intensity of the incident radiation:

$$I_s = A_1 \alpha, \tag{10}$$

where A_1 is the instrument constant, α is the instrument reading.

The radiant flux can be calculated from the relation

$$q_r = \pi I A_w - A_w \sigma_0 T_w^4. \tag{11}$$

If it is permissible to consider the incident radiation isotropic, then

$$I_s = I.$$

In the case of anisotropic radiation it is necessary to find the directivity pattern of the incident radiation and to determine the value of I from the expression

$$I = \frac{1}{2\pi} \int\limits_{2\pi} I_s d\omega;$$

A_w and T_w are determined by direct measurement.

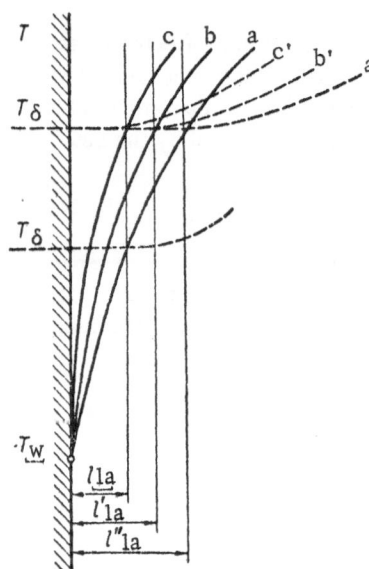

Variation in temperature of the medium at a heat-sensitive surface. a–c) Molecular temperatures; a'–c') radiant temperatures; l_{eq}, l'_{eq}, and l''_{eq} are the distances from the wall to the radiative equilibrium layers.

The convective component can be obtained as the difference between the total heat flux and the radiative component. Placement of the instrument disrupts the continuity of the wall and hence the flow hydrodynamics, so that error can occur in determining the radiative component. When the instrument is placed in a tube its combustion products must not enter. Otherwise distortions of the actual results may ensue.

5. Measurement of the Temperature and Position of the Radiative Equilibrium Layer

The radiative component of heat transfer can be calculated in the case of isotropic incident radiation, if the temperature of the radiative equilibrium layer is known, from the relation

$$q_r = \frac{\sigma_0}{\dfrac{1}{A_w} - \dfrac{1}{2}} \left(T_\delta^4 - T_w^4 \right), \tag{12}$$

where T_δ is the temperature of the radiative equilibrium layer, °K.

The temperature position of the radiative equilibrium layer can be determined by means of a special instrument. Such an instrument is described and the results of measurements given in [2, 7, 13].

The convective component of heat transfer is calculated from the measured (by means of the same instrument) total heat flux q_1:

$$q_\kappa = q_1 - \frac{\sigma_0}{\dfrac{1}{A_w} - \dfrac{1}{2}} \left(T_\delta^4 - T_w^4 \right).$$

(13)

The quantities q_1, A_w, T_δ, and T_w are obtained by direct measurements.

From the value determined for T_δ it is possible to calculate the value of I_δ or E_{in} (assuming isotropic incident radiation) from the expression

$$E_{in.} = \pi I_\delta = \left(T_\delta^4 - \frac{1}{2} A_w T_w^4 \right) \frac{2\sigma_0}{2 - A_w}.$$

(14)

The radiative component for other values of the surface emissivity (e.g., for a test section) can be calculated from the known values of E_{in} and I_δ according to the relation (9).

We wish to analyze the above methods of separating composite heat transfer on the basis of radiative equilibrium layer concepts [7]. We will assume that the equilibrium layer is located at a distance from the heat-sensitive surface of the same order of magnitude as the photon mean free path. Then the radiant and molecular temperature will vary as shown in the figure. The radiant temperature of the equilibrium layer is equal to its molecular temperature. In the layer of thickness l_{eq} next to the wall the radiative energy does not interact with the wall.

In applying the method of two radiometers, with the same incident radiation and different surface emissivities, the temperature of the radiative equilibrium layer must also be different. The temperature of the radiative equilibrium layer, appearing as the radiant temperature, characterizes the radiant energy density. With a decrease in A_w for the same value of E_{in}, the reflected radiant flux and, consequently, the radiant energy density and value of T_δ increase.

In case the variation in radiant flux does not change the molecular temperature field, as A_w is diminished the radiative equilibrium layer must shift slightly away from the wall. In this case it will be located at a distance l'_{eq}. As the emissivity of the wall is increased the equilibrium layer approaches the heat-sensitive surface. If the process is continued according to the indicated scheme, i.e., if the molecular temperature field is not altered, the two-radiometer method will give valid results. If it is assumed that the equilibrium layer is situated in the previous position (curve c) when the surface emissivity of the radiometer is varied, in this case the molecular temperature field and, consequently, the convective component of the heat transfer will change. The two-radiometer method in this case will yield error. With simultaneous variation of the molecular temperature field and distance l_{eq} (curve b) the error will be somewhat smaller than in the foregoing case.

The existing literature on the application of the two-radiometer method does not permit a definite decision regarding the postulates advanced above. Additional detailed research is needed in this area.

In measuring the radiation output we cannot utilize the radiative equilibrium layer concept because in this case the reflected radiation will be practically equal to zero ($A_w = 1$ by agreement). For this case Eqs. (12) and (14) will not be meaningful. In order to obtain a reliable result it would be desirable to conduct simultaneous separation of the composite heat transfer by more than one of the methods outlined above and comparison of the results obtained thereby.

Literature Cited

1. S. S. Filimonov, B. A. Khrustalev, and I. P. Kolchenogova, Experimental Investigation of Heat Exchange in Combustion Chambers, Teploénergetika, No. 7 (1955).
2. L. A. Goryainov, Investigation of Composite Heat Transfer in a Cooled Channel, Tr. LIIZhTa (Transactions of the Leningrad Railroad Engineers Institute), No. 160 (Transzheldorizdat, 1958).

3. G. N. Delyagin and B. V. Kantorovich, Combustion and Heat Transfer of Sprayed Liquid Fuel in Air Eddies, Inzh. -Fiz. Zh., No. 3 (1958).

4. V. I. Lebedev, Effect of the Emissivity of the Furnace Medium on Heat Transfer in Combustion Chambers, Tr. MIITa (Transactions of the Moscow Railroad Engineers Institute), No. 139 (Transzheldorizdat, 1961).

5. S. M. Pokrovskii, Composite Heat Transfer in Liquid Fuel Combustion Chambers, Tr. MIITa, No. 125 (Transzheldorizdat, 1960).

6. B. D. Katsnel'son and A. A. Shatil', Investigation of Heat Transfer in a Horizontal Cyclone Combustion Chamber with Air Cooling, Énergomashinostroenie, No. 11 (1959).

7. P. K. Konakov, S. S. Filimonov, and B. A. Khrustalev, Heat Transfer in Steam Boiler Combustion Chambers [in Russian] (Rechizdat, 1960).

8. É. G. Narezhnyi, Investigation of Heat Transfer in a Gas Turbine Combustion Chamber with a Cool-Air Eddy Chamber, Sudostroenie, No. 10 (1957).

9. I. P. Kolchenogova and S. N. Shorin, Intensification of Heat Transfer in the Combustion of a Gas, Gazovaya Promyshlennost', No. 2 (1959).

10. L. N. Il'in, Effect of Temperature Conditions on Heat Emission and Friction in Airflow in a Tube, Kotloturbostroenie, No. 1 (1951).

11. S. Kocho, Investigation of Heat Transfer in the Working Volume of an Open-Hearth Furnace, Stal, No. 3 (1950).

12. V. N. Adrianov and S. N. Shorin, Heat Transfer in a Flow of Emitting Combustion Products in a Channel, Teploénergetika, No. 3 (1957).

13. P. K. Konakov, S. S. Filimonov, and B. A. Khrustalev, Calculation of Radiative Transfer in Cooled Combustion Chambers, Zh. Tekhn. Fiz., 27(5):(1957).

14. S. V. Ryzhkov, Investigation of Convective Heat Transfer in a Unit Cylinder in a Hot Mazut Flame, Tr. Nikol. Korabl. Inst. (Transactions of the Nikolaev Shipbuilding Institute), No. 15 (1958).

15. S. S. Filimonov, B. A. Khrustalev, and V. N. Adrianov, Investigation of the Convective and Radiative Components of Composite Heat Transfer by the Two-Radiometer Method, Coll.: Convective and Radiative Heat Transfer (Izd. Akad. Nauk SSSR, ÉNIN, 1961).

16. S. S. Filimonov, B. A. Khrustalev, and V. N. Adrianov, Theoretical Foundations of the Two-Radiometer Method, Zh. Tekhn. Fiz., 30(6):(1960).

17. V. N. Adrianov, Radiometric Instrument for the Measurement of Radiant Flows, Coll.: Heat Transfer Problems (Izd. Akad. Nauk SSSR, ÉNIN, 1959).

18. V. T. Kumskov and L. A. Goryainov, Laws of Composite Heat Transfer, Tr. MIITa, No. 125 (Transzheldorizdat, 1960).

CALCULATION OF HEAT TRANSFER IN BOILER FURNACES
BY THE NORMATIVE METHOD

L. A. Goryainov

In 1962 and 1963 changes were introduced into the normative method of calculating the heat transfer in furnaces [1, 2]. We will briefly analyze these changes. The fundamental analytical equation is stated in its earlier form:

$$\theta''_F = \frac{T''_F}{T_a} = \frac{Bo^{0.6}}{Aa_F^{0.6} + Bo_F^{0.6}} \tag{1}$$

or

$$\theta''_F = \frac{\left(\dfrac{Bo}{a_F}\right)^{0.6}}{A + \left(\dfrac{Bo}{c_F}\right)^{0.6}}, \tag{2}$$

where T''_t is the temperature at the furnace exit, °K, T_a is the theoretical combustion temperature, °K, a_t is the emissivity of the furnace, A is a constant coefficient, Bo is the Boltzmann number.

According to the recommendations of the Central Scientific Research Institute for Boilers and Turbines (TsKTI) and the All-Union Heat Engineering Institute (VTI), in every case A = 0.445, except for the flameless combustion of a gas. The dependence of A on the relative height of the burner in the combustion chamber is given in [2] for pulverized coal and gas-mazut furnaces. The values of the contamination coefficients ξ are altered considerably. Since this coefficient is included in the Boltzmann number, the numerical values of the latter will also be modified appreciably from the older recommendations.

The formulas for determining the emissivity of the flame a_f are altered. The coefficient β linking the emissivity of the medium a and that of the flame a_f is excluded from the normative method:

$$a_f = \beta a. \tag{2}$$

This represents a step forward. The unjustifiability and contradiction of the values of the coefficient β were subjected to earlier criticism in [4]. It is recommended that the flame emissivity be calculated according to the formula

$$a_f = ma_{1u} + (1 - m)a_g, \tag{3}$$

where a_{1u} and a_g are the emissivities of luminous and nonluminous gas particles in the flame respectively, and m is a coefficient depending on the type and method of fuel combustion.

The treatment of flame luminosity has also been changed. For m = 0 the flame is nonluminous, for m = 1 it is perfectly luminous; intermediate values are also given for m, equal to 0.2 for the combustion of a gas by a luminous flame and 0.4 for a mazut flame with

$$\frac{Q}{V} = (175 \cdot 10^3 - 233 \cdot 10^3) \ W/m^3.$$

No recommendations are given for choosing the value of m for mazut furnaces in the load interval

$$(233 \cdot 10^3 - 1163 \cdot 10^3) \ \text{w/m}^3.$$

Analytical values.	
Old recommendation	New recommendation
$\psi = 0.969$	$\psi = 0.969$
$T_a = 2131°\text{K}$	$T_a = 2131°\text{K}$
$\xi = 0.7$	$\xi = 0.45$
$\text{Bo} = 0.432$	$\text{Bo} = 0.671$
$\beta = 0.65$	—
$a_f = \beta a = 0.65 \cdot 1 = 0.65$	$a_f = 0.8$
$a_F = 0.60$	$a_F = 0.901$
$\left(\dfrac{\text{Bo}}{a_F}\right)^{0.6} = 0.825$	$\left(\dfrac{\text{Bo}}{a_F}\right)^{0.6} = 0.84$
$\theta'' = 0.65$	$\theta'' = 0.654$

The coefficient of 0.82, which previously represented the absorptive power of the radiation-sensitive heating surface, is omitted from the formulas for determining the furnace emissivity. This has no conceptual significance.

The analytical equation uses the ratio Bo/a_F rather than a_F; therefore the omission of the coefficient 0.82 from the formula for a_F is actually compensated by the new values of the contamination coefficient. As tests have shown, the emissivity of the screened heat-sensitive surface is nearer to 1 than to 0.82.

The changes effected in certain quantities by the new recommendations are illustrated in a sample calculation, borrowed from [1], of the heat transfer in a pulverized coal boiler furnace.

Consequently, in calculating the heat transfer in furnaces by the new recommendations, despite the rather significant changes in a_F and Bo, the basic empirical equation remains the same. Clearly, this can be accomplished by a suitable choice of contamination coefficients and the quantities appearing in the formula for a_F. It would hardly be possible, therefore, for any of the coefficients to have any strictly defined physical meaning.

We will conduct an analysis of the effect of individual factors on the final result of calculations using Eq. (1). Invoking the method of error theory, we differentiate Eq. (1) with respect to Bo, a_F, and A. After suitable transformations, we obtain

$$\frac{d\theta}{\theta} = \frac{1}{1 + \dfrac{1}{A}\left(\dfrac{\text{Bo}}{a_F}\right)^{0.6}} \left[0.6\,\frac{d\text{Bo}}{\text{Bo}} - 0.6\,\frac{da_F}{a_F} - \frac{dA}{A} \right] \qquad (4)$$

or

$$\frac{d\theta}{\theta} = \frac{1}{1 + \dfrac{1}{A}\left(\dfrac{\text{Bo}}{a_F}\right)^{0.6}} \left[0.6\,\frac{d\left(\dfrac{\text{Bo}}{a_F}\right)}{\dfrac{\text{Bo}}{a_F}} - \frac{dA}{A} \right],$$

where

$$\frac{d\theta}{\theta} = \frac{dT_2}{T_2}.$$

For the example treated above, we determine the effect of a change in the individual quantities on the final result. We let A = 0.445, whereupon

$$\frac{d\theta}{\theta} = 0.21 \left[\frac{d\text{Bo}}{\text{Bo}} - \frac{da_F}{a_{F'}} \right] - 0.35\,\frac{dA}{A}.$$

As the value of Bo is increased the relative influence on θ, dBo/Bo, da_F/a_F, and dA/A is diminished. We investigate the expression for $a_F = a_f/[a_f - (1 - a_f)\psi\xi]$ in analogous fashion. We obtain

$$\frac{da_t}{a_t} = \frac{(1 - a_f)\,\xi\psi}{a_f + (1 - a_f)\,\xi\psi}\ \frac{d\psi}{\psi}.\tag{5}$$

In order to determine the influence of a change in the degree of screening ψ on θ, we substitute into Eq. (4) the value of da_F/a_F from Eq. (5), letting $dBo = 0$ and $dA = 0$, whereupon we get

$$\frac{d\theta}{\theta} = -\,0.6\ \frac{1}{1 + \dfrac{1}{A}\left(\dfrac{Bo}{a_F}\right)^{0.6}}\ \frac{(1 - a_f)\,\xi\psi}{a_f + (1 - a_f)\xi\psi}\ \frac{d\psi}{\psi}.\tag{6}$$

The minus sign in Eq. (6) indicates that the furnace exit temperature drops as the screening is increased. For our example

$$\frac{da_F}{a_F} = 0.0983\,\frac{d\psi}{\psi} \quad \text{and} \quad \frac{d\theta}{\theta} = -0.0206\,\frac{d\psi}{\psi}.$$

Let us consider the effect of a change in ψ on θ for a mazut furnace with $s \geq 2.5$ m. We assume $\psi = 0.9$, $\xi = 0.6$, and that the ratio Bo/a_F is the same as in the example of the pulverized coal furnace, so that $da_F/a_F = 0.0567\ d\psi/\psi$ and $d\theta/\theta = -0.0119\ d\psi/\psi$.

For $d\psi/\psi$ of the order 0.4 the value of θ will change by about 0.5%. For $T_2 = 1300°K$ this yields a change in temperature of the furnace exit of about 6.5°K. For large values of Bo the change in temperature will be still smaller.

Consequently, our analysis of Eq. (1) confirms for large furnaces the postulate advanced in [3], that the degree of screening does not affect the heat transfer in mazut furnaces.

The constant value of a_f used in the analytical procedure for pulverized coal and mazut furnaces at $s \geq 2.5$ m shows that the temperature at the furnace exit depends only on one constant coefficient characterizing the kind of fuel and on the two variables Bo and ψ. In this part, after the correction in number of initial variables has been introduced, the normative method agrees with the method of P. K. Konakov [4]. Konakov's method, however, has deeper physical meaning than the normative method, the analytical equation of this method being derived from the thermal balance equation. A comparison of the experimental data with the results of calculations according to both methods shows that the Konakov calculation yields better agreement with experiment by comparison with the normative method.

The above analysis of the variations in calculation of heat transfer in boiler furnaces by the normative method permits the assertion that the added corrections are inadequate, since the fundamental analytical equation, left in its old form, is empirical and contradicts the law of conservation of energy, while the individual quantities appearing in it are insufficiently justified from the physical point of view. For example, in cyclone burners θ can be greater than 0.9, so that Eq. (1) cannot possibly be used for the analysis of cyclone burners.

Literature Cited

1.   Thermal Calculation of Boiler Assemblies; Normative Method (Gosénergoizdat, 1957).
2.   L. M. Gurvich and V. V. Mitor, Calculation of Heat Transfer in Gas-Mazut and Pulverized Coal Furnaces, Énergomashinostroenie, No. 2 (1963).
3.   P. K. Konakov, S. S. Filimonov, and B. A. Khrustalev, Heat Transfer in Steam Boiler Combustion Chambers (Rechizdat, 1960).
4.   P. K. Konakov, Theoretical Foundations of Heat Engineering (Transzheldorizdat, 1957).

EXPERIMENTAL DETERMINATION OF EFFECTIVE
FLAME EMISSIVITY BY MEANS OF A DIRECTIONAL RADIOMETER

S. M. Pokrovskii and V. I. Lebedev

The calculation of heat transfer in a furnace chamber by the standard method using norms [1] requires a determination of the furnace emissivity, for which empirical formulas and nomograms are used. Recently the authors of the norms have published a number of modifications and corrections to the procedure for calculating heat transfer in furnace chambers [2].

The present article gives the results of an experimental determination of the emissivity of the flame in the combustion of liquid fuel (kerosene) and gas (propane) in pure form and with the addition of chrome–magnesium dust, as well as a comparison with the analytical data obtained using the equations recommended in [1] and [2].

The experimental determination of the flame emissivity is based on relations prescribed by the theory of radiative heat transfer. The specific radiant flux incident on a heat-sensitive surface located in the furnace chamber is equal to

$$E = \varepsilon_f \sigma_0 T^4, \tag{1}$$

where ε_f is the emissivity of the flame, σ_0 is the black body radiation coefficient, T is the mean molecular temperature of the emitting medium.

In the case of black body radiation, for which $\varepsilon_{BB} = 1$,

$$E_{BB} = \sigma_0 T_{BB}^4. \tag{2}$$

Equating the expressions (1) and (2) for identical specific radiant fluxes, we obtain

$$\varepsilon_f = \left(\frac{T_{BB}}{T} \right)^4. \tag{3}$$

In engineering calculations of radiative heat transfer it may be assumed with sufficient accuracy that

$$\varepsilon_f = a_f.$$

To measure the radiant flux we used a directional radiometer, which is described in detail in [3]. The radiometer was placed in windows located in the wall of the chamber. The radiation from the flame was allowed to impinge on the thermosensitive element of the radiometer, consisting of a pile of Chromel–Copel thermocouples.

Prior to the experiments the radiometer was calibrated against the radiation from a black body model in the form of a platinum oven with a small iris aperture. The calibration data were used to construct a characteristic of $E_{BB} = f(T_{BB})^4$. The temperature T_{BB} in the black body model was measured with a platinum–platinorhodium thermocouple.

In the furnace chamber tests we measured the efficiency E_t of the radiometer thermopile, a quantity proportional to the incident radiant flux, and the temperature of the furnace medium in the cross section of the chamber where the radiometer was placed. The temperature of the medium was measured with a suction type platinum–platinorhodium thermocouple. The mean temperature derived from 12 measurements of the local temperature values at different points in the direction of radiation was used in the calculations.

S. M. POKROVSKII AND V. I. LEBEDEV

TABLE 1

Test No.	Temperature, °K furnace medium chamber section			Temperature, °K black body chamber section			Flame emissivity, ε_f chamber section			Mean values of ε_f	T_2, °K	$\frac{Q}{V}$ W/m³	Air excess ratio, a_a	Values of a_f obtained by methods recommended in references: [2]	[1]
	I	II	III	I	II	III	I	II	III						
1	1221	1135	1021	1152	1100	993	0.792	0.882	0.845	0.833	741	291	1.86	0.318	0.142
2	1282	1185	1048	1230	1162	1012	0.845	0.921	0.870	0.877	939	530	1.95	0.262	0.195
3	1373	1178	1043	1268	1157	1032	0.730	0.922	0.940	0.863	964	557	2.03	0.270	0.202
4	1418	1234	1105	1240	1162	1032	0.758	0.810	0.758	0.773	1013	645	1.86	0.278	0.214
5	1383	1178	1055	1172	1120	993	0.520	0.810	0.750	0.695	942	584	2.08	0.266	0.195
6	1323	1173	1046	1180	1135	1012	0.630	0.882	0.882	0.797	941	593	2.03	0.260	0.195
7	1330	1153	1043	1200	1120	1012	0.662	0.885	0.883	0.810	973	610	2.06	0.273	0.206
8	1408	1223	1096	1250	1180	1063	0.610	0.862	0.883	0.785	1043	658	1.71	0.290	0.225
10	1446	1226	1133	1240	1175	1047	0.540	0.845	0,730	0.700	1057	738	1.64	0.291	2.226
11	1383	1183	1081	1223	1157	993	0.608	0.900	0.705	0.738	1022	635	1.88	0.283	0.218
13	1228	1123	1018	1090	1012	943	0.622	0.655	0.730	0.667	902	426	2.15	0.284	0.187
16	1366	1348	1205	1355	1340	1180	0.955	0.930	0.918	0.934	1128	1110	2.92	0.295	0.246
17	1415	1331	1141	1220	1200	1135	0.545	0.655	0.97	0.723	1149	1045	2.64	0.298	0.248
18	1472	1305	1086	1265	1195	1062	0.55	0.695	0.9	0.714	1133	968	2.46	0.296	0.296

TABLE 2

Mass flow of gas G_g, kg/sec	Air excess ratio, a_g	Temperature, °K				Flame emissivity, ε_f		Mean values of ε_f	T_s, °K	Value of a_f obtained by method recommended in [1]
		furnace medium chamber section		black body chamber section		chamber section				
		I	II	I	II	I	II			
0.00258	1.26	1320	1285	1251	1215	0.207	0.219	0.217	1235	0.125
0.00258	1.26	1250	1232	1192	1168	0.606	0.640	0.623	1200	0.310*
0.00347	1.28	1227	1200	1100	1081	0.600	0.628	0.615	1196	0.305*
0.00320	1.30	1283	1238	1215	1193	0.198	0.212	0.205	1211	0.124
0.00347	1.21	1237	1201	1143	1119	0.200	0.220	0.210	1300	0.126
0.00347	1.21	1300	1253	1227	1200	0.617	0.653	0.635	1251	0.325*
0.00363	1.15	1372	1335	1301	1271	0.218	0.224	0.221	1324	0.130
0.00363	1.15	1309	1271	1252	1235	0.615	0.675	0.645	1272	0.335*
0.00347	1.22	1351	1308	1273	1244	0.180	0.221	0.201	1305	0.120
0.00347	1.22	1293	1241	1219	1163	0.600	0.636	0.618	1257	0.323*
0.00448	1.10	1450	1415	1342	1328	0.195	0.221	0.208	1390	0.132
0.00448	1.10	1382	1336	1318	1279	0.651	0.710	0.681	1330	0.347*

* Concentration of chrome–magnesium dust in the furnace medium, 0.2 kg/nm³ exhaust gases.

The flame emissivity in the combustion of liquid fuel was measured in a horizontal cylindrical furnace chamber 950 mm long with an inside diameter of 400 mm. Furnace screens of copper tubing 12 mm in diameter were attached to the walls of the chamber. The degree of screening $\psi = 0.335$. The radiation output was measured in three equally separated cross sections of the tube.

In the gas experiments the length of the chamber was 950 mm; the inside diameter was 330 mm. A solid cylindrical screen was placed on the walls of the furnace. The degree of screening of the chamber was $\psi = 0.413$. The radiation output was measured in two cross sections.

The results of the measurements and calculations for liquid fuel are shown in Table 1, those for the gas in Table 2.

For the investigated furnace chamber we calculated the values of a_f using the equations and nomograms recommended in [1] and [2]. The calculated values of a_f and the experimental values of ε_f are shown in Tables 1 and 2. Because of the lack of adequate recommendations for the value of the coefficient m at the existing furnace loads, its value was obtained by interpolation. A comparison of the results shows that calculating the flame emissivity according to the recommendations given in [1] and [2] yields values that are too small for ε_f. It is evident, therefore, that the coefficient a_f in the proposed method of calculation [1] does not have physical meaning but is an empirical coefficient whose function is to compensate for the physical heat transfer process unaccounted for in the analytical procedure, specifically, the heat transfer due to convection.

Literature Cited

1. Thermal Calculation of Boiler Assemblies; Normative Method (Gosénergoizdat, 1957).
2. A. M. Gurvich and V. V. Mitor, Calculation of Heat Transfer in Gas-Mazut and Pulverized-Coal Furnaces, Énergomashinostroenie, No. 2 (1963).
3. P. K. Konakov, S. S. Filimonov, and B. A. Khrustalev, Heat Transfer in Steam Boiler Combustion Chambers (Rechizdat, 1960).

COMPARISON OF EXPERIMENTAL DATA WITH COMPOSITE HEAT TRANSFER CALCULATIONS IN BOILER FURNACES BY THE NORMATIVE METHOD AND THE MIIT METHOD

V. I. Lebedev and S. M. Pokrovskii

The present article presents a comparison between the experimental data and furnace exit temperatures calculated by the normative method [1], including the corrections to that method as indicated in [3], and by the method of the Moscow Institute of Railroad Engineers (MIIT) [2].

We investigated the heat transfer in the furnace of a single-drum boiler with a capacity of 32 (metric) tons per hour and steam parameters of 43 atm and 440°C, as well as in a type BKZ-210-140-f boiler furnace.

The single-drum boiler assembly utilizes forced recirculation of water in the evaporative heating surfaces. The fuel, natural gas, was compressed in a chamber kiln.

The volume of the furnace chamber was 145 m^3, the effective radiative heating surface 140 m^2, the degree of screening $\psi = 0.76$. The normal burner output (two front burners and two side burners) was 750 nm^3/h. The gas pressure in front of the burners was regulated with a slide valve, the air pressure by a damper placed in the air line ahead of the burner. The feed water was supplied in a water economizer heated to 100°C. Complete balanced testing of the boiler system was carried out in the furnace in our investigation of heat transfer. The flow of gaseous fuel was accounted for by a throttle device (diaphragm) placed in the gas line and connected to a float type differential manometer with two secondary indicating and recording instruments. The amount of gas consumed during the test period was determined by planimetry of the recording flowmeter disk diagram. The results were compared with the data of the indicating instruments. The results of the measurements were generally in good agreement.

The composition of the compressed gas and its heat of combustion were determined throughout the experiment by laboratory sampling.

The quantity of steam developed in the test period was ascertained from the built-in reading and recording instruments, as well as a diaphragm placed in the steam line.

The pressure and temperature of the superheated steam were measured with a manometer and Chromel-Copel thermocouple connected to a recording potentiometer and indicating logometer.

The boiler system was tested with closed drains and without blowing, enabling us to monitor the steam production of the boiler assembly against the flow of feed water. In this case the water level in the boiler drum remained unchanged at the beginning and end of the test. The flow of feed water was measured with throttle devices. The pressure and temperature of the water fed to the boiler were also measured in the feed line.

The temperature of the combustion products in the exit cross section of the furnace was measured at five points with a suction type platinum–platinorhodium thermocouple.

An analysis of the gases in the exit cross section was performed with a VTI (All-Union Heat Engineering Institute) gas analyzer. The air heating temperature was measured, along with all other measurements required for setting up the thermal balance of the boiler assembly. A total of 18 tests were run on the boiler assembly.

TABLE 1

Test No.	Fuel rate, B, nm³/sec	Combustion heat of gas, Q_n^I, kJ/nm³	Air excess ratio, α	Heat loss from incomplete chemical combus., q_3, %	Air preheat temperature, t_a, °K	Available heat in furnace, Q_r, kJ/nm³	Composition of combustion products in furnace exhaust, %						Theoretical combustion temperature, T_1, °K	Furnace exit temperature, °K		
							CO_2	O_2	CO	H_2	CH_4	$C_n H_m$	T_1	T_2	T'_2	T''_2
1	0.481	24300	1.11	2.92	410	23700	9.66	8.70	0.0	0.0	0.3	0.0	1833	1094	1043	1061
2	0.500	23600	1.32	1.35	416	23300	8.24	5.35	0.05	0.0	0.1	0.0	1593	1083	1047	1061
3	0.672	22900	1.16	2.01	416	23000	9.34	3.26	0.0	0.0	0.2	0.0	1738	1201	1131	1145
4	0.997	22000	1.08	3.89	424	21700	9.70	2.46	0.1	0.0	0.38	0.0	1701	1263	1194	1211
5	1.040	21400	1.12	1.32	424	21600	9.62	2.60	0.1	0.0	0.1	0.0	2041	1293	1255	1269
6	0.473	23800	1.41	1.23	418	24100	7.74	6.32	0.0	0.0	0.1	0.0	1893	1100	1047	1080
7	0.454	24000	1.23	1.07	408	24600	8.91	4.20	0.0	0.0	0.1	0.0	2021	1067	1013	1023
8	0.883	24000	1.16	0.00	422	24600	9.72	4.00	0.0	0.0	0.0	0.0	2273	1301	1237	1255
9	0.810	26800	1.34	1.89	423	26900	10.64	3.37	0.0	0.0	0.2	0.0	2033	1275	1212	1235
10	0.623	27000	1.42	1.16	420	27400	8.21	5.52	0.0	0.0	0.1	0.0	2017	1199	1146	1165
11	0.625	25800	1.09	1.97	414	26000	9.60	3.01	0.0	0.5	0.4	0.0	2199	1206	1136	1175
12	0.720	28300	1.13	5.10	424	29300	9.80	2.9	0.9	0.0	0.36	0.0	2133	1277	3380	1258
13	0.753	28300	1.15	5.25	423	29200	10.30	1.7	0.6	0.0	0.50	0.0	2113	1291	1231	1273
14	0.760	28400	1.07	3.54	432	29300	9.0	4.0	0.1	0.3	0.30	0.0	2248	1271	1205	1238
15	0.734	24400	1.23	3.74	434	29800	9.8	2.4	0.0	0.0	0.20	0.0	2083	1285	1213	1258
16	0.720	28500	1.05	2.63	426	30200	8.6	4.8	0.3	0.3	0.10	0.0	2293	1258	1156	1215
17	0.652	28200	1.15	3.16	424	30000	9.5	1.7	0.3	0.3	0.20	0.0	2145	1242	1163	1191
18	0.623					30000	9.5	3.5	0.0	0.4				1226		1204

TABLE 2

Test No.	Fuel rate, nm^2/sec	Fuel combustion heat, kJ/kg	Air excess ratio	Heat into furnace, kJ/kg	Heat into furnace, kJ/kg			Theoretical combustion temperature, °K	Furnace exit temperature, °K		
									heat balance	norms	MIIT method
	B	Q_n^r	α	Q_F	q_3	q_4	q_5	T_1	T_2	T_2'	T_2'
1	3.85	40000	1.25	43700	0.26	—	0.6	2246	1283	1361	1325
2	7.02	40000	1.10	43200	1.50	—	0.6	2428	1343	1360	1362
3	2.46	45000	1.05	41800	0.17	0.01	0.6	2449	1344	1190	1249
4	2.91	40000	1.22	44200	0.42	0.01	0.6	2305	1302	1290	1296
5	2.50	40000	1.27	43700	0.45	0.01	0.6	2218	1226	1220	1241
6	2.84	39700	1.08	42400	0.31	0.01	0.6	2426	1278	1249	1288
7	2.62	40000	1.15	43300	0.87	0.01	0.6	2360	1287	1230	1275
8	2.44	39400	1.23	43800	0.33	0.01	0.6	2277	1243	1229	1251
9	3.95	40000	1.07	43600	0.14	—	0.6	2498	1454	1382	1328
10	3.84	40000	1.09	43700	0.13	—	0.6	2469	1412	1370	1384
11	2.86	40000	1.17	43900	0.42	—	0.6	2357	1275	1275	1292
12	3.00	39200	1.23	44200	0.49	—	0.6	2290	1309	1300	1298
13	2.00	40000	1.44	44900	0.61	—	0.6	2058	1187	1188	1182
14	2.07	39900	1.40	44800	0.89	0.02	0.6	2104	1185	1192	1190
15	3.50	40000	1.05	43500	0.87	—	0.6	2530	1439	1330	1378
16	3.61	40400	1.04	43100	0.15	—	0.6	2523	1441	1339	1382
17	2.93	40400	1.29	44200	0.87	—	0.6	2219	1295	1295	1288
18	3.56	40000	1.08	43300	0.84	0.13	0.6	2462	1326	1330	1352
19	3.57	40000	1.09	43000	2.00	0.02	0.6	2442	1346	1340	1360
20	3.37	40000	1.15	43700	5.25	0.01	0.6	2381	1365	1330	1345
21	3.36	40000	1.14	43500	0.14	0.01	0.6	2387	1313	1330	1343

The tests on the BKZ-210-140-4 boiler were carried out through the efforts of VTI. The results of the tests have been published in [6]. The boiler had a chamber kiln for the combustion of mazut. The total wall surface of the furnace chamber was 679.2 m². The radiating surface of the open smooth-tube screens amounted to 508.2 m². Twenty-one tests were run on the boiler. The discrepancies in heat balance during the tests were between 2 and 4%. The temperature obtained at the furnace exit from the heat balance equation was noted for comparison.

In processing the experimental data according to the MIIT method, the analytical coefficient z appearing in the heat balance equation of [2] was used. The value of this coefficient was taken from the results of composite heat transfer investigations in furnace chambers operating with gas [4] and liquid fuel [5].

Table 1 gives the results of the measurements and calculations of the temperatures at the exit cross section according to the normative method and the MIIT method for the single-drum boiler. Table 2 gives the same data for the BKZ-210-140-f boiler.

It is apparent from the tables that the most satisfactory agreements is obtained in the MIIT calculations.

The mean relative error in determining the furnace exit temperature in the case of the single-drum boiler was 5.1% in the normative method, 2.2% in the MIIT method; the same values for the BKZ-210-140-f boiler were 3.94% and 2.4%, respectively.

It should also be realized that the MIIT method is based on the law of conservation of energy, has fewer empirical coefficients, and requires less time to carry out the calculations.

All of this lends justification to the MIIT method as the preferred basis for heat transfer calculations in furnace chambers.

Literature Cited

1.   Thermal Calculation of Boiler Assemblies; Normative Method (Gosénergoizdat, 1957).
2.   P. K. Konakov, S. S. Filimonov, and B. A. Khrustalev, Heat Transfer in Steam Boiler Combustion
      Chambers (Rechizdat, 1960).
3.   A. M. Gurvich and V. V. Mitor, Calculation of Heat Transfer in Gas-Mazut and Pulverized Coal Furnaces,
      Énergomashinostroenie, No. 2 (1963).
4.   V. I. Lebedev, Investigation of the Effect of Emissivity of the Medium on Heat Transfer in a Furnace
      Chamber, Inzh.-Fiz. Zh., No. 12 (1960).
5.   S. M. Pokrovskii, Calculation of Liquid Fuel Boiler Furnace Chambers, Inzh.-Fiz. Zh., No. 12 (1962).
6.   A. D. Gorbanenko et al., Combustion of Liquid Fuel in Furnace Chambers, Teploénergetika, No. 4 (1963).

RAISING THE EFFICIENCY OF ANTHRACITE FURNACES

V. G. Erokhin and A. I. Pokalyuk

When the forced draft in a furnace is increased there occurs a considerable increase in the gas flow velocity, which in turn raises the thermal losses due to incomplete combustion of a chemical and mechanical nature. Lengthening the trajectory of the gas flow and swirling it inside the furnace chamber increases the time of incumbency of the furnace gases and exhaust gases in the furnace, which promotes more thorough mixing of the combustion products with air and more complete burning. The result is a decrease in the thermal losses q_3 and q_4. This is accomplished by a variety of measures, one of which is the so-called "fine jet" technique. The fine jet is created by injecting small (fine) jets of air, steam, smoke gases, or a mixture of these through special nozzles located at a certain height above the fuel bed.

The fine jets emerge at high velocities and penetrate the thickness of the furnace gases, swirling them and mixing them vigorously with the oxygen in the air. This promotes more complete combustion of the active gases and ignition of the volatile coke particles, reducing the losses q_3 and q_4.

Theoretically, in order to create turbulence of the gas flow in the furnace chamber it is possible to use any type of fine jet: air, steam, gas, air and steam, or gas and steam. In practice the most widely used type in low-power boiler equipment is air and steam, whereas, according to published information, gas jets have yet to be employed.

Nevertheless, this type of jet has certain advantages over the air jet, especially in anthracite furnaces, in which better cooling of the grate is achieved with gas jets.

We conducted an experimental program for the purpose of investigating the effects of gas jets on the rated efficiency of a furnace and of comparing with other types of jets.

The object of the tests was an anthracite coal furnace used to fire a type A-7 Shukhov-Berlin boiler. The furnace was equipped with a fine-jet assembly consisting of a high-head ventilator fan, steam ejector, collector with nozzles and ancillary lines. A unidirectional jet system with frontal placement of the nozzles was used. The calculations for the setup were carried out according to regular procedure. The tests were conducted with sorted anthracite coal, of type AK. The operation of the furnace was investigated without the jets and with air, gas, and gas–steam jets.

The tests were conducted after the so-called "inverse balance" method, i.e., based on a determination of the thermal losses and efficiency of the furnaces as a residual balance term.

In order to ascertain the advantages and disadvantages of the investigated types of jets relative to furnace operation without jets, it was assumed that the operating efficiency of the furnace depended on the magnitude of the thermal losses with incomplete combustion of chemical and mechanical origin.

The heat loss with the exhaust gases is only partially dependent on the efficient operation of the furnace, since it is caused by the amount of air excess.

Consequently, in order to compare the efficiency of operation of the furnace with a particular type of jet, we used the so-called rated efficiency of the furnace:

$$\eta_f^{rat.} = 100 - (q_2' + q_3 + \Sigma q_4) \%,$$

where q'_2 is the heat loss with the exhaust gases, determined at a temperature difference $t_{ex} - t_{in} = 200°C$ [VTI (All-Union Heat Engineering Institute) method], %; q_3 and Σq_4 are the heat losses due to chemical and mechanical (total) unburned products, %.

The heat loss q_5 to the surrounding medium of the furnace is not included in the rated efficiency, since it is assumed to be identical in operation of the furnace with or without any of the various types of jet.

The losses q'_2 and q_3 were determined from the test data on the basis of the conventional equations, and the losses Σq_4 were determined by the VTI method for manually stoked furnaces.

For the air-jet tests the air was tapped by the ventilator from the boiler compartment. The air pressure in the collector was 180 mm H_2O, the mean air flow velocity on emerging from the nozzles was 48.8 m/sec.

The gas and steam jet was created by the ejection of furnace gases with saturated steam ($p_{ej} = 5.3$ atm). The gases were pumped from the second gas conduit of the boiler with a mean experimental temperature of 530°C. The pressure of the gas-stream mixture in the collector was 162 mm H_2O at its mean experimental temperature of 244°C; the mean velocity of the gas–steam jets at the nozzle exit was 40 m/sec.

The gas jet was realized by circulating the exhaust gases with the ventilator fan at $t_{ex} = 187°C$. The nozzle velocity of the gases was 84.5 m/sec.

As is apparent from the accompanying table summarizing the results of the tests, the rated efficiency of the furnace with all of the jet types turned out to be 4-5% higher than the furnace efficiency without a fine jet. The rated efficiency of the furnace was increased due to a reduction in the principal furnace losses q_3 and Σq_4, which is a result of more complete mixing of the hot gases and entrainment particles with oxygen when the fine jets are used.

The reduction in thermal loss due to incomplete chemical combustion was a maximum by comparison with other types of jet in the case of gas jets, for which, in spite of the greater air excess in the furnace ($\alpha_f = 2.14$), the thermal loss q_3 was 1.78%, as opposed to 7.2% for the furnace without fine jets.

In analyzing the components q_4 of the thermal loss due to incomplete burning of a mechanical nature, we notice the effective reduction in q_4^{en} (entrainment) in cases when the furnace was operating with fine jets. The reduction in q_4^{en} was greatest for air jets, in which case q_4^{en} amounted to 1.41%, against $q_4^{en} = 4.76\%$ for the furnace operating without fine jets.

In addition to the reduction in furnaces losses q_3 and q_4 in cases of furnace operation with fine jets, there occurred an increase

Test	Q_f^u (from Mendeleev form.), kJ/kg	α_f	α_{ex}	V_0, nm³/kg	V_2, nm³/kg	c_{com}, kJ/mm³	q_4^{oth} %	q_4^{sl} %	q_4^{en} %	Σq_4 %	CO_r	Q'_2, kJ/kg	q'_2	Q'_3, kJ/kg	q_5	η_f
Without jets	28030	1.52	1.87	6.52	11.49	1.33	0.3	1.6	4.76	6.66	1.43	2620	9.3	2010	7.2	76.8
With air jet	27500	1.87	2.12	6.9	13.78	1.34	0.3	1.68	1.41	3.4	0.77	3130	11.4	1280	4.65	80.55
With steam-gas jet	26600	1.87	2.12	6.34	13.425	1.34	0.3	2.05	1.72	4.08	0.523	3140	11.9	830	3.14	81.3
With gas jet	27300	2.14	2.39	6.9	16.1	1.335	0.3	1.8	1.73	3.83	0.308	3770	13.9	484	1.78	80.5

in the normalized heat loss with the exhaust gases q_2'. This was a consequence of the increases air excess in the furnace when operating with the fine jets. We note in this case that the increased excesses of air can be obviated by suitable regulation of the jets.

It is evident from a comparison of the test results with the various kinds of fine jet that the highest rated efficiency was obtained when a steam-gas jet was used.

In evaluating the types of jet investigated above, the gross rated efficiency of the furnace was taken into account, whereas the economy of efficiency of the jet technique must be evaluated in terms of the net efficiency. The latter takes into account the flow of energy for a particular type of fine jet according to intrinsic need: in the case of a ventilator jet about 1 or 2% of the boiler steam capacity, in the case of a steam-gas (ejector type) jet 0.6-0.7%.

As the tests showed, the use of air and gas jets, as well as steam-gas jets, is fully justified, even with the combustion of sorted anthracite in the furnace and manual stoking, and is one way of enhancing the efficiency and lowering the cost of the boiler system.

Conclusions

1. All three types of fine jet (air, gas ventilator, and steam-gas) yield almost the same thermal effect.

2. By comparison with the operation of a furnace without fine jets, the application of the three indicated types of fine jet yields a 4-5% increase in the rated efficiency of the furnace for combustion of sorted anthracite, grade AK.

3. The practical design of the fine jet and its utilization in the USSR and elsewhere has demonstrated that the realization of very low-capacity (about 6-8% of the total air flow for combustion) but high-head ventilators (300-400 mm H_2O) involves sizable difficulties, particularly in low-power boiler systems. However, the ejector type steam-gas jet makes it possible to get along without high-head fans, while in addition it is possible to tap smoke gases from any gas line of the boiler, i.e., at any gas temperature, for ejector type steam-gas jets. These two factors constitute the prime advantages of the steam-gas ejector jet over the ventilator fan version.

4. The thermal effect gained from the use of all three types of fine jet, i.e., the 4-5% increase in rated furnace efficiency, is typical of furnace equipment operating with sorted grades of fuel.

5. Particular stress is placed on the gas jet in comparison with the air jet in the burning of unsorted anthracite, grade ARSh, which is primarily the fuel used in low- and even in medium-power boiler sets. The injection of inert gases into the furnace chamber as a working medium for the jet not only eliminates the increase in mean air excess α_f throughout the entire combustion process, it also greatly enhances the efficiency of the furnace due to more effective diminution of the thermal loss q_4^{en}, which is the prevalent type of loss in the burning of unsorted anthracite over all other thermal losses of a mechanical nature (slag and other):

$$(q_4^{sl} + q_4^{oth}).$$

Consequently, the gas jet is best used in cases when it is particularly urgent to cope with incomplete burning due to mechanical factors, i.e., when the fuel has a high content of fine fractions (ARSh and ASSh coals) and crater type combustion tends to occur.

Literature Cited

1. R. G. Granovskii, The Fine Jet in Furnace Equipment (Gosénergoizdat, 1947).

CONVECTIVE HEAT TRANSFER IN THE TURBULENT MOTION OF A VISCOUS FLUID IN THE INITIAL SECTION OF A TUBE

V. S. Sidorov

Convective heat transfer in the motion of a fluid in the initial section of a cylindrical channel is of considerable practical importance. Normally, in determining the coefficient of convective heat transfer in the initial section (L/d < 50), a correction factor is introduced on the basis of the data given in [10]. There are no analytical relations for convective heat transfer that are suitable for practical application. It is reasonable that attempts should have been made in certain papers to investigate the problem analytically [3, 8]. However, the derived relations are poorly confirmed by experiment.

In the present article we analyze convective heat transfer in the initial section of a cylindrical tube and, on the basis of integral relations, obtain a solution that conforms nicely to the experimental data.

In an isothermal fluid flow inside a cylindrical channel a boundary layer is formed, its thickness being a function of the distance from the entrant cross section, i.e., $\delta = \psi(z)$. Within the limits of the boundary layer the flow velocity w varies from 0 at the wall to a value w_s in the unperturbed portion of the flow (free stream) (Fig. 1). At a distance z from the entrant cross section we cut out an element dz and examine the variation in momentum of the fluid in the boundary layer of this element.

The following momentum is gained by the fluid as it passes through the cross section I of the boundary layer:

$$2\pi \int_0^\delta \rho (r_0 - y)\, w^2 dy.$$

In the section II the momentum gained changes by an amount

$$2\pi dz \frac{d}{dz} \int_0^\delta \rho (r_0 - y)\, w^2 dy.$$

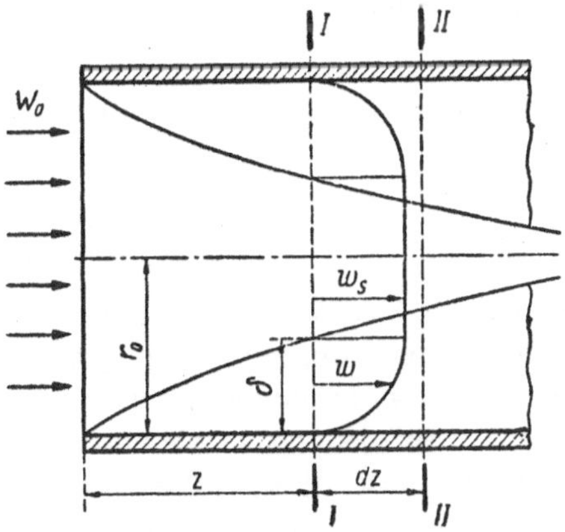

Fig. 1. Diagram of fluid motion in the cylindrical channel.

Since the amount of matter entering through the inner surface of the element from the unperturbed part of the turbulent flow is eight times the amount of matter leaving through the same surface, and the difference between them is equal to [1]

$$2\pi dz \frac{d}{dz} \int_0^\delta \rho (r_0 - y)\, w dy,$$

it may be assumed that the total change in momentum of the medium through the inner surface of the element is equal to

$$2\pi w_s dz \frac{d}{dz} \int_0^\delta \rho (r_0 - y)\, w dy.$$

75

According to the second law of mechanics, the change in momentum is equal to the impulse due to the forces acting on the element dz.

We express the forces due to pressure and viscous (fluid) friction acting on this element. In the section I the pressure force acting on the element is

$$\pi\delta\,(2r_0 - \delta)\,p,$$

and in the section II the force is

$$\pi\,(\delta + d\delta)\,[2r_0 - (\delta + d\delta)]\,(p + dp).$$

If we assume that the total mass flow of fluid through any cross section of the tube is constant, the velocity w_s will increase from w_0 in the entrant cross section to w_{sec} after the initial section. Consequently, from the direction of the unperturbed stream the boundary layer will be acted upon by a pressure force

$$2\pi\,(r_0 - \delta)\,pd\delta.$$

Computing the projections of all the pressure forces on the z axis, we find that the element is acted upon by a pressure force

$$-\,2\pi r_0\delta dz\,\frac{dp}{dz}\,.$$

If we let τ_0 denote the specific friction force of the fluid against the wall, the total viscous friction force is expressed as

$$-\,2\pi r_0\tau_0 dz.$$

Determining the pressure gradient dp/dz from the Bernoulli equation and assuming that the flow density ρ varies only insignificantly, we equate the total change in momentum of the fluid to the impulse of the external forces acting on the system, after which we obtain the equation for the conservation of momentum of the boundary layer fluid in the form

$$w_s\left[\delta\frac{dw_s}{dz}+\frac{d}{dz}\int_0^\delta\left(1-\frac{y}{r_0}\right)wdy\right]-\frac{d}{dz}\int_0^\delta\left(1-\frac{y}{r_0}\right)w^2dy=\frac{\tau_0}{\rho}\,. \tag{1}$$

Equation (1) is valid not only for fluid motion in a tube, but over a plate as well. In this case we must let $r_0 = \infty$, w_s = const.

For turbulent motion with Re ranging up to 10^5, it may be assumed that the cross-sectional velocity distribution obeys the law

$$\frac{w}{w_s}=\left(\frac{y}{\delta}\right)^{1/7}. \tag{2}$$

From the condition of flow continuity we determine the velocity function $w_s = \psi(\delta)$. Bearing in mind that in the turbulent motion of a fluid the thickness of the laminar substrate does not exceed 0.02 d [2], the velocity being almost equal to zero therein, it is permissible (for a definite mass flow of fluid through the cross section) to extend the assumed velocity distribution law (2) right up to the wall without appreciable error. It then becomes possible to express the free stream velocity w_s in terms of the fluid velocity w_0 in the entrant cross section of the channel. We find by integration of Eq. (2) that w_{av} = 0.82 w_s. Then we can assert that the following momentum is imparted to the boundary layer:

$$1.64\,\pi\rho r_0\left(1-\frac{\delta}{2r_0}\right)w_s\delta, \tag{3}$$

and the unperturbed flow is imparted a momentum of

$$\pi \rho r_0^2 \left(1 - \frac{\delta}{r_0}\right)^2 w_s. \tag{4}$$

Inasmuch as the amount of fluid passing through the entrant cross section of the tube is

$$\pi \rho r_0^2 w_0, \tag{5}$$

it follows that, equating the sum of the values of (3) and (4) to the value of (5) and denoting

$$1.64 \frac{\delta}{r_0} \left(1 - \frac{\delta}{2r_0}\right) + \left(1 - \frac{\delta}{r_0}\right)^2 = A,$$

we obtain the following expression for the velocity w_s:

$$w_s = \frac{w_0}{A}. \tag{6}$$

Adopting the special Prandtl hypothesis of the mixing length [5] or proceeding from the Blasius law of frictional resistance [3], we obtain an expression for τ_0:

$$\tau_0 = 0.0228 \rho w_s^2 \left(\frac{\nu}{w_s \delta}\right)^{0.25}. \tag{7}$$

Relying on Eqs. (2), the velocity distribution law, (6), the value of the flow velocity in the z direction, and (7), the friction force, the equation of conservation of momentum of the boundary layer can be reduced to a first-order differential equation:

$$f(\delta) \, d\delta = \frac{0.027}{\mathrm{Re}^{0.25}} \, dz, \tag{8}$$

where

$$f(\delta) = \left(\frac{\delta}{r_0}\right)^{0.25} \times \frac{2.8A\left(\frac{7}{72} - \frac{7\delta}{120r_0}\right) + \left(\frac{23}{72} + \frac{98}{240}\frac{\delta}{r_0}\right)\frac{\delta}{r_0}\left(1 - \frac{\delta}{r_0}\right)}{2.8A^{1.25}},$$

$$\mathrm{Re} = \frac{w_0 d}{\nu}.$$

Equation (8) cannot be solved in the form represented, owing to the complexity of the function $f(\delta)$. Analysis has shown that the function $f(\delta)$ is a parabola, with a maximum value equal to 0.11 at $\delta/r_0 \approx 0.55$. The equation of this parabola may be written in the form

$$0.11 - f(\delta) = \left|\frac{\delta}{r_0} - 0.55\right|^n.$$

Determining the exponent n from the boundary condition $f(\delta)|_{\delta/r_0 = 1} = 0.4$, we arrive at n = 3.33. Then

$$f(\delta) = 0.11 - \left|\frac{\delta}{r_0} - 0.55\right|^{3.33}. \tag{9}$$

Making use of the resultant expression (9), we integrate Eq. (8). Once the constant of integration has been ascertained from the condition $|\delta|_{z=0} = 0$, we can obtain the solution to Eq. (8):

$$4.07 \frac{\delta}{r_0} - \frac{\left|\dfrac{\delta}{r_0} - 0.55\right|^{4.33}}{0.117} + 0.633 = \frac{1}{\mathrm{Re}^{0.25}} \cdot \frac{z}{r_0}. \tag{10}$$

For practical purposes the derived relation may be written in the form

$$\frac{\delta}{r_0} = \frac{0.15}{\mathrm{Re}^{0.312}} \left(\frac{z}{r_0}\right)^{1.25}.$$ (11)

An analysis of the relation (11) shows that for $\delta = r_0$ the stabilization length is

$$L = 4.43 \, \mathrm{Re}^{0.25} r_0.$$

A comparison of the experimental data with Latzko's solution [4] and with Eq. (11) (Fig. 2) shows that Eq. (11) agrees far better with the experimental data than Latzko's solution.

Using the expression (11) for the variation in thickness of the boundary layer along the length of the tube we can derive analytical relations for the convective heat transfer in the initial section. In this case the motion of the fluid will be nonisothermal in the initial section. According to the well-known Reynolds analogy in the theory of heat transfer, we have

$$\frac{\mathrm{Nu}}{\mathrm{PrRe}} = \frac{\tau_0}{\left[1 + \frac{w_l}{w_s}(\mathrm{Pr}-1)\right] w_s^2 \rho},$$ (12)

where w_l is the velocity of the fluid at the upper boundary of the laminar substrate.

According to the data obtained in [2],

$$\frac{w_l}{w_s} = 1.88 \left(\frac{w_s \delta}{\nu}\right)^{-0.125}.$$ (13)

With this expression and Eqs. (6) and (7) we can reduce Eq. (12) to the form

$$\frac{1}{\mathrm{Nu}} = \frac{A^{0.75} \left(\frac{\delta}{r_0}\right)^{0.125}}{0.027 \, \mathrm{Re}^{0.75} \mathrm{Pr}} \left[\left(\frac{\delta}{r_0}\right)^{0.125} + \frac{2.05 A^{0.125}(\mathrm{Pr}-1)}{\mathrm{Re}^{0.125}}\right].$$ (14)

An estimate of the term $A^{0.125}$ as δ/r_0 varies from 0 to 1 shows that it varies from 0.98 to 1.0. The sec-

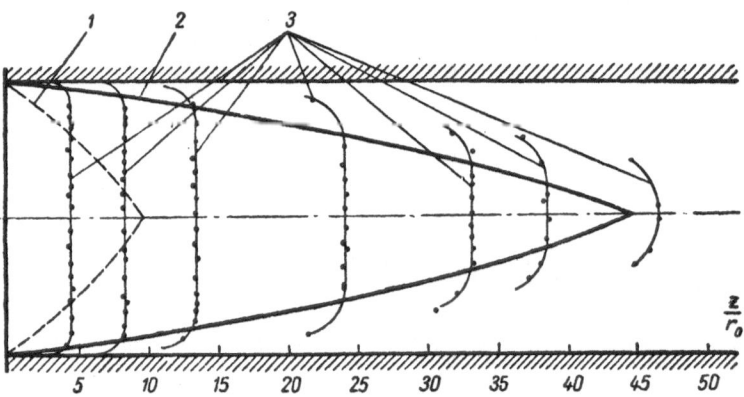

Fig. 2. Comparison of experimental data with Lazko's solution and with Eq. (11). 1) Latzko's solution; 2) solution of Eq. (11); 3) experimental data.

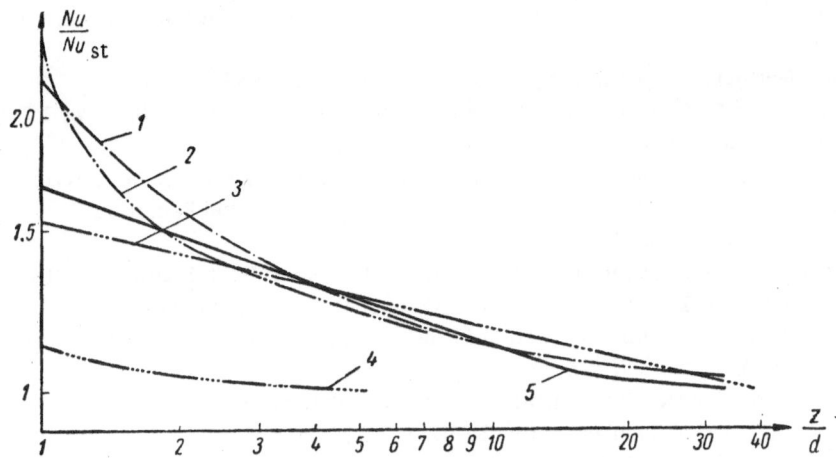

Fig. 3. Comparison of the analytical equation (17) with data of other investigations. 1) Experiments of Filimonov and Khrustalev; 2) experiments of Grass; 3) experiments of Alad'ev; 4) according to Deissler's theory; 5) according to Eq. (17).

ond term in the brackets may therefore be rewritten approximately as

$$\frac{2.05 A^{0.125}\,(\text{Pr} - 1)}{\text{Re}^{0.125}} \approx \frac{2.025\,(\text{Pr} - 1)}{\text{Re}^{0.125}} = C,$$

whereupon for the thermodynamic stabilization length at $\delta = r_0$ we obtain Eq. (14) in the form

$$\frac{1}{\text{Nu}_{st}} = \frac{1 + C}{0.0313\,\text{Pr}\text{Re}^{0.75}}; \tag{15}$$

for gases with $\text{Pr} \approx 1$ Eq. (15) is brought to the form

$$\text{Nu}_{st}^2 = 0.0313\,\text{Re}^{0.75}. \tag{16}$$

Equation (16) is in almost complete agreement with Mikheev's formula [9]. Taking the ratio of Eq. (15) to (14), we obtain a correction for calculating the heat transfer in the initial section as follows:

$$\frac{\text{Nu}}{\text{Nu}_{st}} = \frac{0.863\,(1 + C)}{A^{0.75}\left(\dfrac{\delta}{r_0}\right)^{0.125}\left[\left(\dfrac{\delta}{r_0}\right)^{0.125} + C\right]}. \tag{17}$$

As in the first case, the resultant expression (17) is compared with the results of experiments (Fig. 3). It is clear that, beginning with $z/d > 2$, the solution agrees nicely with experiment. The divergence in the interval $0 < z/d < 2$ is explained by the assumption made in this article that the turbulent layer is formed immediately at the beginning of the tube. In reality, however, a laminar boundary layer is formed near the entrant cross section with turbulent motion, the laminar layer going over to turbulent flow after a certain finite distance. Because the thickness of the laminar boundary layer is somewhat less than the turbulent boundary layer, the ratio Nu/Nu_{st} has to increase somewhat more rapidly in this interval.

Literature Cited

1. P. K. Konakov, S. S. Filimonov, and B. A. Khrustalev, Heat Transfer in Steam Boiler Combustion Chambers (Rechizdat, 1960).

2. E. R. G. Eckert and R. M. Drake, Theory of Heat and Mass Transfer [Russian translation] (Gosénergoizdat, 1961).

3. H. Schlichting, Boundary Layer Theory [Russian translation] (IL, 1956).

4. L. Schiller, Motion of Fluids in Circular Tubes [Russian translation] (ONTI, Moscow, 1936).

5. L. Prandtl, Hydro- and Aeromechanics [Russian translation] (IL, 1949).

6. S. S. Filimonov and B. A. Khrustalev, Calculation of Heat Transfer and Viscous Friction in the Turbulent Motion in Tubes with Various Entry Conditions, Coll.: Convective and Radiative Heat Transfer [in Russian] (Izd. Akad. Nauk SSSR, 1960).

7. I. T. Alad'ev, Experimental Determination of Local and Average Heat Transfer Coefficients for Turbulent Fluid Flow in Tubes, Izv. Akad. Nauk SSSR, Otd. Tekhn. Nauk, No. 11 (1954).

8. Deisler, Turbulent Heat Transfer and Friction in the Entrance Regions of Smooth Passages, Trans. ASME, Vol. 77, No. 8 (1955).

9. M. A. Mikheev, Heat Transfer Fundamentals (Gosénergoizdat, 1956).

10. A Short Engineering Physics Handbook, Vol. 3 (Fizmatgiz, 1962).

EXPERIMENTAL INVESTIGATION OF CONVECTIVE HEAT TRANSFER
AND FLUID FRICTION IN THE MOTION OF A VISCOUS FLUID
WITH LARGE REYNOLDS NUMBERS

K. F. Aksenov, V. V. Kudryavtsev, Yu. A. Lomov,
A. I. Pokalyuk, and Yu. N. Khvoshchevskii

It is generally known that the dimensionless group called the Reynolds criterion or Reynolds number is expressed in the form

$$\text{Re} = \frac{wd_e}{\nu}, \tag{1}$$

where w is the flow velocity of the medium, m/sec; d_e is the equivalent channel diameter, m; ν is the kinematic viscosity coefficient of the medium, m^2/sec.

The Reynolds number is one of the fundamental criteria governing the hydrodynamic similarity of fluid flows.

As implied by Eq. (1), the Re number can be increased in geometrically defined channels, on the one hand, by increasing the linear flow velocity in the channel and, on the other, by decreasing the kinematic viscosity of the medium.

The first approach creates the familiar problem of heat transfer and viscous (fluid) friction in the motion of a medium at high velocities.

The Reynolds number can also be written as follows:

$$\text{Re} = \frac{\rho w d_e}{\mu}, \tag{2}$$

where ρ is the density of the medium, kg/m^3, μ is the dynamic viscosity coefficient, N · sec/ m^2.

As apparent from Eq. (2), the Re number and the flow velocity w can be increased at the same time by also increasing the density of the medium ρ.

For air and other diatomic gases at pressures up to 100 atm, the equation of state of an ideal gas is valid with a sufficient degree of accuracy for practical purposes:

$$\frac{p}{\rho} = RT.$$

Consequently, the expression for Re can be rewritten as

$$\text{Re} = \frac{wpd_e}{RT\mu}. \tag{3}$$

The dynamic viscosity coefficient μ does increase with pressure, but only very slightly. For example, with an increase in p from $9.8 \cdot 10^4$ N/m^2 at t = 0°C, μ increases 15%, at t = 150°C it increases 2%. Consequently, at constant velocity and temperature of flow it may be considered to a first approximation that the Re number varies in proportion to the absolute pressure p.

The Reynolds number can also be increased by lowering the absolute temperature of the flow. However, for gas heat transfer agents used in industrial heat exchanges this method does not suffice to produce any significant change in the Re number and is not of practical interest. In this regard, the temperatures of the heat transfer agents in heat exchangers are normally prescribed.

In the design of heat exchanger equipment the amount of flow of the heat transfer agent is usually given, so that it becomes convenient to express the Re number in terms of the mass flow of the medium G in kg/sec:

$$Re = \frac{Gd_e}{F\mu},$$

where F is the cross-sectional area of the channel, m^2.

Since

$$d_e = \frac{4F}{f},$$

it follows that

$$Re = \frac{4G}{f\mu},$$

where f is the perimeter of the channel cross section in contact with the fluid, m.

For a circular tube

$$Re = \frac{4G}{\pi d\mu}. \tag{4}$$

Consequently, for each specific channel the Reynolds number is determined primarily by the mass flow of the given heat transfer agent.

The existing relations between similarity criteria for convective heat transfer and viscous friction, of the form

$$Nu = \varphi(Re, Pr), \tag{5}$$

and

$$\zeta = \psi(Re, Pr) \tag{6}$$

do not reflect the effects of pressure. Since the Pr number and coefficient of heat conductivity of the gases λ are changed very little with increasing pressure, the analytical heat transfer coefficient calculated from these relations for a given Re number will be almost the same, regardless of how this number evolves, whether by the flow velocity or by the pressure.

The experimental investigations on which the relations (5) and (6) are based were conducted for the most part at atmospheric pressure or just slightly above it, so that it would not be possible to verify the admissibility of extrapolating the analytical formulas for the heat transfer and viscous friction to high pressures.

In the majority of instances the indicated analytical equations are also used for the calculations appropriate to heat exchanging equipment operating at rather high pressures, for example, steam superheaters for boilers. This is really only justified, however, by the absence of more reliable formulas, plus the fact that the resultant heat transfer coefficient in the steam superheater depends to a large degree on the coefficient of heat transfer from the gases to the tube walls α_2, rather than on the coefficient of heat transfer from the tube walls to the steam flow α_2, so that even a fairly large error in determining the heat transfer coefficient α_2 only slightly affects the resultant heat transfer coefficient k.

The study of heat transfer and viscous friction at high pressures has been retarded not only by the ominous engineering difficulties in carrying out the necessary experiments, but also by the fact that until very recently there have not been sufficiently reliable data on the physical parameters of heat transfer agents at high pressures.

The investigation of convective heat transfer associated with the motion of an incompressible fluid (droplets) has shown that the pressure value does not significantly alter the heat transfer rate.

The research that has been conducted thus far on heat transfer and viscous friction at high Re numbers for compressible fluids covers only a very small range of pressures, although certainly it is of utmost practical interest to establish analytical relations for the heat transfer and viscous friction of gas flows at the high Re numbers that can be realized at high pressures. It is known that the pressure can be increased by reducing considerably the weight and dimensions of heat exchangers and decreasing the expenditure of energy for circulation.

It is normally assumed at the present time that the conventional dimensionless criterion relations for convection heat transfer, as obtained in experiments at atmospheric pressure, are also valid at the high pressures realized in heat transfer agents.

As far back as 1909, while conducting experiments on the heating of air during motion in tubes at pressures from $19.6 \cdot 10^4$ to $137.2 \cdot 10^4$ N/m^2 (in an interval of numbers Re = 3000 to 200,000), Nusselt [1] established the fact that within the given limits of pressure variation the heat transfer coefficient is determined solely by the mass flow rate of air $u = w_\rho$, independently of the separate factors involved in this product, i.e., the velocity w and medium density ρ.

In 1939 Nefedov [2] investigated experimentally the convective heat transfer in a flow of mixed nitrogen and hydrogen (ratio 1 : 3) in a tube at pressures from $9.8 \cdot 10^4$ to $2705 \cdot 10^4$ N/m^2. The experiments were conducted with a smooth steel tube 7.66/18 mm in diameter and 2 m in length. The velocities of the gas were varied from 2 to 9 m/sec, corresponding to a variation in Reynolds number from 2700 to 190,000. The experimental tube was heated with steam. The mean temperature of the gas in the tests amounted to 50–60°C.

The experiments confirmed the validity of the dimensionless criterion relation

$$Nu = \varphi \, (Re, \; Pr, \; l/d).$$

The experimental data obtained by Nefedov on convective heat transfer are satisfactorily described by the following equation for the entire range of indicated pressures and Re numbers:

$$Nu = 0.0212 \, Re^{0.8}, \; Pr^{0.37}. \tag{7}$$

Similar experiments were conducted in 1947 by the American researchers Colburn, Drew, and Worthington [3]. They also investigated heat transfer in the motion of a hydrogen–nitrogen mixture (3 : 1) in a tube at pressures from $29.4 \cdot 10^4$ to $8820 \cdot 10^4$ N/m^2. The tests were conducted with a steel tube 16 × 38 mm in diameter, 1.5 m in length. The range of Re numbers covered by the experiments was 40,000 to 440,000.

Moreover, using the same apparatus, tests were conducted on the heat transfer in air flowing in a tube at a pressure of $49 \cdot 10^4$ to $78.4 \cdot 10^4$ N/m^2 and Re numbers = 9000 to 50,000. The experiments corroborated the possibility of generalizing the experimental data on heat transfer to a single formula. Although the authors did not process their experimental results in the form of dimensionless criteria, the experimental points corresponding to the various pressures, constructed from the results of these experiments, provided a satisfactory fit to one straight line in logarithmic coordinates according to the equation

$$\frac{a}{c_p} = \; \psi(u_s),$$

where c_p is the specific heat of the gas, u_s is the mass velocity of the gas.

Experiments on heat transfer in free convection in nitrogen at a pressure of $9800 \cdot 10^4$ N/m^2 also confirmed the applicability of the conventional criteria relations, in the form

$$Nu = \varphi \, (Gr, \; Pr),$$

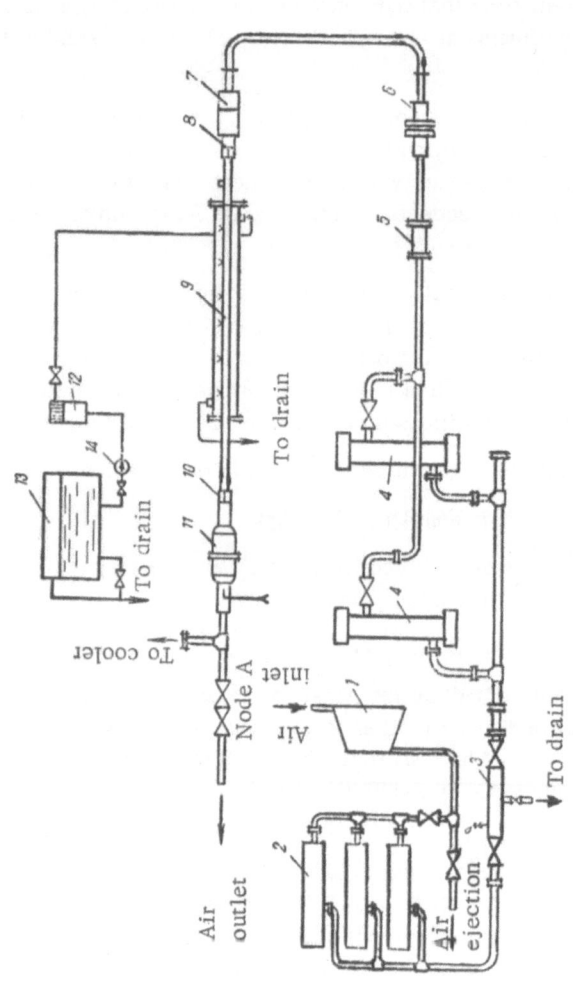

Fig. 1. Diagram of experimental arrangement. 1) Compressor; 2) oil separator; 3) receiver; 4) electric oven; 5) mixer; 6) flow nipple; 7) measurement chamber at inlet to test section; 8) anterior fitting; 9) working section; 10) posterior fitting; 11) measurement chamber at outlet from test section; 12) rotameter; 13) flow tank; 14) water pump.

where

$$\mathrm{Gr} = \frac{g d^3 \beta \Delta T}{\nu^2}$$

is the Grashof number.

A number of authors are of the opinion, however, that extrapolation of the experimental relation on heat transfer as derived from the results of investigations with equipment operating at atmospheric pressure is not justifiable in application to large Reynolds numbers.

In 1953 and 1954 an investigation was conducted on the heat transfer and aerodynamic friction in banks of smooth tubes at Reynolds numbers in excess of 100,000 [4]. The tests, conducted in a wind tunnel, showed that at an air pressure of $9.8 \cdot 10^4$ N/m^2 (Re = $20 \cdot 10^3$ to $100 \cdot 10^3$) the experimental data fully satisfy the well-known analytical heat transfer relations of the All-Union Heat Engineering Institute (VTI) and Central Scientific Research Institute for Boilers and Turbines (TsKTI). The interval of larger Re numbers, however, obtained at the expense of an increase in flow pressure as indicated above, is typified by a substantial increase in heat transfer rate. Consequently, the TsKTI, VTI, and other formulas become inapplicable for calculations for Re > 120,000. In the criteria relation Nu = CRe^n the exponent of Re increases to 0.79 for in-line tube banks and to 0.94 for staggered banks.

These tests also demonstrated the inapplicability of the existing relations for determining the aerodynamic friction in tube banks when Re $\geq$ 300,000.

The lack of sufficiently reliable experimental data on convective heat transfer and viscous friction at high Reynolds numbers like those realized at high pressures creates considerable problems in calculations of heat exchangers with high heat transfer parameters.

The objective of our work was to investigate experimentally the laws of convective heat transfer and viscous (fluid) friction at air pressure with high Reynolds numbers in cooled channels, as well as to study the influence of coolant pressure on the convective heat transfer and viscous friction.

Experimental Arrangement

The tests were run on an experimental apparatus (Fig. 1) designed and built especially for the investigation of convective heat transfer and viscous friction at air pressure in cooled channels with high Reynolds numbers. Air at a pressure up to $686 \cdot 10^4$ N/m^2 was forced by a type 5VP 16/70 stationary piston-operated four-stage compressor with a nominal capacity of 16 nm^3/min into three parallel-connected oil separators. Each oil separator was a thick-walled vessel with built-in separating devices (Fig. 2). The air received through the through the inlet pipe passed through an attachment consisting of annular slotted channels (louvers), where it abruptly altered its direction and was directed through the annular space between the housing and sections to the outlet pipe, located in the middle part of the oil separator housing.

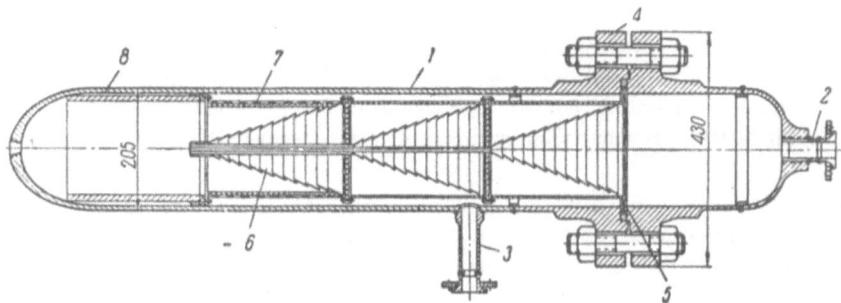

Fig. 2. Oil separator. 1) Housing; 2) inlet pipe; 3) outlet pipe; 4) flanges; 5) metal grids; 6) conical section of louvered separator; 7) cylindrical metal grid; 8) bottom of oil separator.

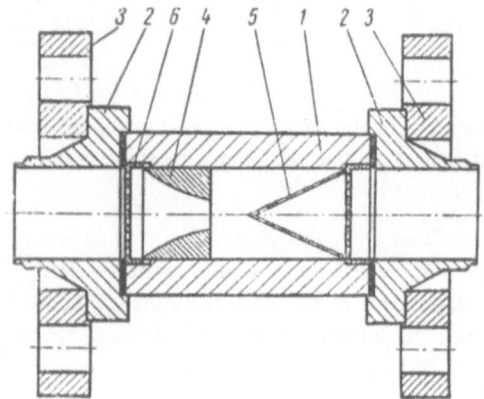

Fig. 3. Mixer. 1) Housing; 2) shoulders; 3) flanges; 4) nozzle; 5) cone; 6) metal grids.

The air was delivered from the oil separators to a receiver with approximately one cubic meter capacity. The receiver consisted of fifteen oxygen tanks connected in parallel by means of connecting pipes to a common collector (steel tube 219 × 19 mm in diameter).

Connecting the receiver into the loop practically eliminated air pulsations from the working section. Furthermore, the receiver acted as a moisture–oil separator, for which the collector was provided with drain cocks. The oil residues in the air leaving the oil separators were heated as they passed through electric air heaters, whose elements had a temperature of the order 600–700°C. The air heaters were tubular electric ovens with a design rating of 65 kW. These ovens were unique in that their heating elements were Nichrome tubes 8 × 1 mm in diameter, through which the air passed in direct contact. The tubes were heated by an electric current at 60 V from three welded single-phase transformers, type TS-500, with current regulators. Heating of the air was regulated by varying the current through the low side of the supply transformers.

In order to achieve maximum temperature equalization of the hot air, the latter was delivered from the ovens to a mixer. The mixer (Fig. 3) was constructed in the form of a nozzle and perforated copper cone pointed against the current, so that a temperature field uniform throughout the cross section of the tube was ensured at the entry to the measurement and experimental sections of the equipment.

After the mixer and in front of the throttle diaphragm was a straight section of tube 50 mm in diameter and about 2.5 m in length.

The test section of the equipment was a water–cooled cylindrical channel, or steel tube 2 m long (Fig. 4). An initial uncooled section about fifty times the diameter of the experimental tube in length was provided for hydrodynamic stabilization of the air flow. The test section was connected into the equipment loop by means of fittings. Bell-and-spigot joints ensured a good fit and rapid changing of the experimental channel. The pressure and air flow were regulated by valves, eight of which were located in the entire loop. The cooling water was supplied from the water line to a feed tank with a capacity of 1.5 m³, located at a height of 4 m above floor level. From the feed tank the water was delivered by an electrically driven centrifugal pump at a pressure of $19.6 \cdot 10^4$ N/m² through a regulating valve and rotameter into the experimental section. In every case concurrent flow of the heat transfer agents (water and air) was used, since this mode provides the most uniform wall temperature on the part of the cooled channel.

In the course of the experiments we measured the air temperature in front of the throttle diaphragm and at the inlet and outlet of the experimental tube section, as well as the wall temperature of the tube and the temperature change of the cooling water. For the measurements we used a Chromel–Copel thermocouple. The temperature of the cooling water was monitored by means of glass mercury thermometers.

For a reliable measurement of the air temperature at the outlet from the experimental section a special measuring chamber was provided, in which six thermocouples had been installed. The hot junctions of the thermocouples (beads) were placed at different points of the cross section at three different radii, ever 60° in circumference.

The diameter of the measurement chamber was considerably larger than the channel diameter, which provided for a reduction in flow velocity to 2–5 m/sec, and a grid placed in front of the set of thermocouples promoted equalization of the velocity and temperature profile in the flow cross section.

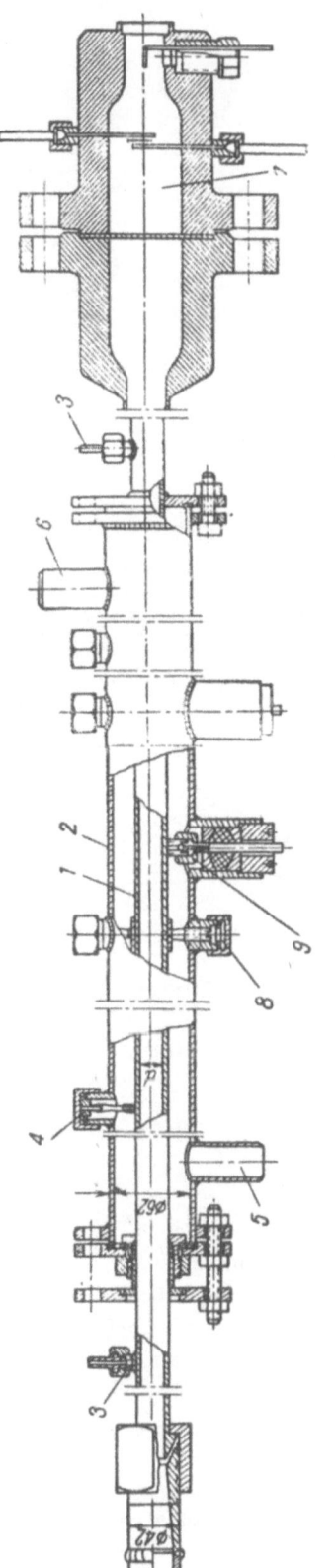

Fig. 4. Diagram of experimental section of test equipment. 1) Experimental tube; 2) casing; 3) bleed pipes for pressure measurements; 4) branch pipe for thermocouple insertion; 5) water inlet; 6) water outlet; 7) measurement chamber; 8) lock; 9) branch pipe for pressure measurements.

For measurement of the wall temperature of the experimental section six annular grooves were cut into the outer surface of the tube. Holes 1 mm in diameter and 0.5 mm deep were drilled in the grooves at different points along a helical line. The beads of Chromel–Copel thermocouples were calked into the wells created thereby. The glass–wool insulated thermoelectrodes were inserted into the grooves and led away through branch pipes welded to the outer surface of the tube.

To measure the emf of the thermocouples we used a low-resistance dc potentiometer, type R306, class 0.015, with a second-class normal saturated element and high-sensitivity portable galvanometer, type M198 magnetoelectric system, with a light indicator. Moreover, the readings of the thermocouples were recorded on a strip chart by means of an electronic automatic-recording potentiometer, type ÉPP-09M1.

We were able by continuous recording of all the measured temperatures every 72 sec to determine the instant at which steady conditions came about and to monitor the heat transfer process throughout the entire test.

The static air pressure, pressure drop at the throttle diaphragm, and head loss due to friction over the length of the experimental section were measured with type DT-150 differential manometers, which had type MKD spring-loaded control manometers on the high-pressure side.

The mass velocity of the air flow was measured by means of a normal throttle diaphragm. The flow of cooling water was measured by means of type RS-5 rotametric flow indicators.

Investigations and Conclusions

More than 50 tests were run on the experimental apparatus with air flowing in smooth tubes of different diameters, along with a series of tests involving the artificial creation of turbulence in the air flow by means of coil turbulators.

The investigated range of Re numbers was from $180 \cdot 10^3$ to $1.45 \cdot 10^6$; the pressure range was $58.8 \cdot 10^4$ to $588 \cdot 10^4$ N/m². The air flow velocity varied from 8 to 90 m/sec.

The tests conducted on convective heat transfer in the motion of air in technically smooth cylindrical cooled channels (tubes) showed that in the tested interval of Reynolds numbers ($0.18 \cdot 10^6$ to $1.45 \cdot 10^6$) the experimental values of the heat transfer coefficients agree satisfactorily with their calculated values according to the criteria formula $Nu = 0.023 \, Re^{0.8} + Pr^{0.4}$ (Fig. 5). Consequently, in order to determine the heat transfer coefficients for air flowing in smooth tubes it is permissible to use the normal criterial relations. The tests also confirmed the postulate that the principal quantity determining the rate of heat transfer is the Re number, regardless of whether it is realized by virtue of the flow velocity or corresponding pressure.

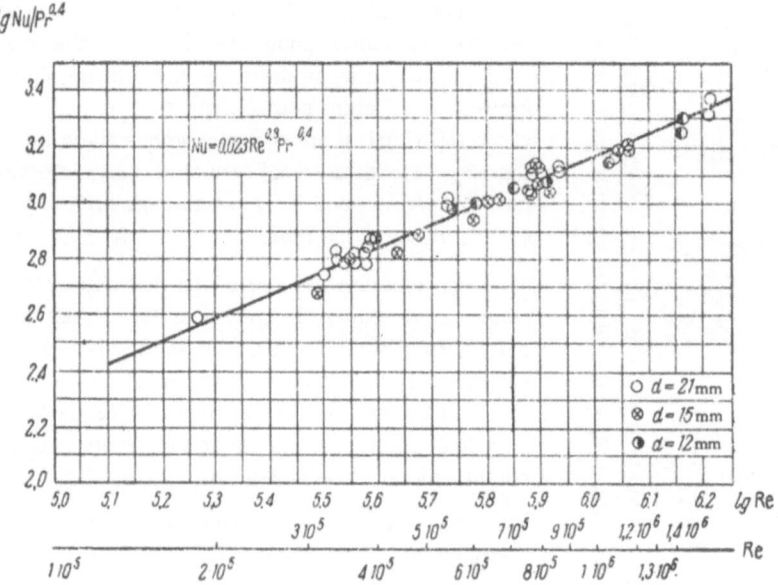

Fig. 5. Graph showing the dependence Nu/Pr$^{0.4}$ = f(Re) in logarithmic coordinates.

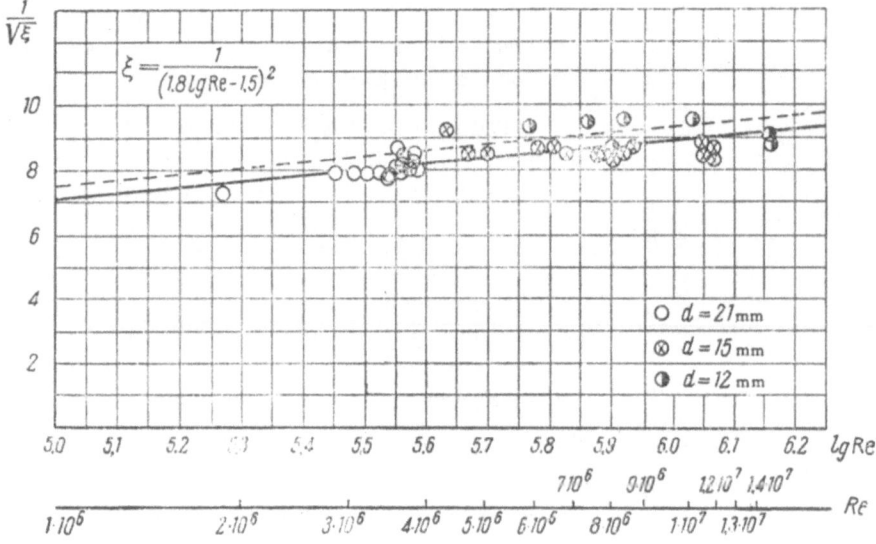

Fig. 6. Dependence of the coefficient of viscous friction on the Reynolds number.

An analysis of the experimental data on viscous friction showed that the dependence of the coefficient of friction differs only slightly from the well-known relations of Prandtl, Nikuradze and Konakov, and other authors for the isothermal motion of incompressible fluids in smooth tubes. Figure 6 gives the function

$$\frac{1}{\sqrt{\zeta}} = f\left(\lg \text{Re}\right)$$

according to the formula of Konakov (dashed line); also plotted on this graph are the experimental points from the data of direct measurements of the pressure drop in the experimental section of the channel. The dependence of the coefficient ε on the Re number indicates that the investigated interval refers to smooth flow conditions. In order to clarify the effect of pressure on convective heat transfer in other than smooth conditions, a series of tests were conducted on the convective heat transfer under conditions of artificial turbulence. The experiments with turbulators showed that under these conditions the convective heat transfer is not only affected by the Re number but also by the pressure or velocity of the heat transfer agent.

Given the same Re but different velocities of the medium, the heat transfer rate does not remain the same. For example, according to the data from tests for the one Re number = 464,000 and an air pressure of 41 atm, the heat transfer coefficient and Nu number were 6 to 8% higher than at a pressure of 26 atm.

For Re $= 1.2 \cdot 10^6$ and pressure increased from 21 to 38 atm the heat transfer coefficient and Nu number increased by about 10–14%, i.e., in the case of high pressure on the part of the heat transfer agent with a given Re (for lower flow velocity) the heat transfer coefficient is higher than at lower pressure. Consequently, the intensification of convective heat transfer by artificial creation of turbulence in the flow of a compressible medium is more effective in the case of small velocities, i.e., at higher pressures on the medium, rather than at higher velocities corresponding to low pressure.

One must be very cautious, therefore, about extrapolating the criteria relations for heat transfer to other pressures when the heat transfer agent is flowing under conditions of artificial turbulence.

It would be advisable in the future to establish specific relations for heat transfer and viscous friction with the artificial creation of turbulence in the flow (for various types of turbulators or swirling devices), taking pressure effects into account.

The analytical formulas for calculating the coefficient of convective heat transfer under these conditions should have the defining dimensionless criteria augmented, in addition to Re and Pr, by the dimensionless group RT/w_0^2 or the Maieevskii–Mach criterion.

Literature Cited

1. W. Z. Nusselt, Verein deutscher Ingenieur (1909). Mutt, Furschungsarbeit, 89:1 (1910).
2. A. B. Nefedov, Heat Transfer in Gases at High Pressure, Zh. Khim. Prom., No. 8 (1939).
3. A. Colburn, F. Drew, and H. Worthington, Heat Transfer in a 3–1 Hydrogen–Nitrogen Mixture at High Pressure, Industrial and Engineering Chemistry, 39(8):(1947).
4. M. F. Lyalin, Heat Transfer and Aerodynamic Friction of Smooth-Tube Clusters for Gas Flows with Large Reynolds Numbers, Teploénergetika, No. 9 (1956).

DETERMINATION OF THE REYNOLDS NUMBER
IN HEAT TRANSFER RELATIONS

L. A. Goryainov, V. A. Beilin, and V. A. Pavlenko

The analytical relations for convective heat transfer and viscous (fluid) friction in the forced motion of a fluid usually involve the Reynolds number, which is the ratio of the inertial forces to the internal friction forces. In the present article we investigate some of the characteristics of various approaches to assessment of the Re number. As is known, the physical parameters contained in the number can be normalized to various defining temperatures. The numerical values of the Re number are calculated according to the formulas

$$\mathrm{Re}' = \frac{wd}{\nu}, \tag{1}$$

$$\mathrm{Re}'' = \frac{Gd}{\mu f}, \tag{2}$$

where w is the flow velocity of the medium, m/sec; d is the defining scale dimension, m; ν is the kinematic viscosity of the medium, m²/sec; μ is the dynamic viscosity of the medium, N · sec/m²; G is the mass velocity of the medium, kg/sec; f is the cross section of the flow channel, m².

In Eq. (1) the calculation is made in terms of the linear velocity, in (2) it is made in terms of the mass flow. These two expressions are not always identical. If the physical parameters are normalized to a temperature other than the mean flow temperature, the values of Re calculated according to Eqs. (1) and (2) will differ.

Let the physical parameters be normalized to a defining temperature T_{def}. Knowing that

$$w = \frac{G}{f\rho_{fl}} \quad \text{and} \quad \nu_{def} = \frac{\mu_{def}}{\rho_{def}},$$

we obtain after substitution of the values of w and ν into Eq. (1),

$$\mathrm{Re}' = \frac{Gd}{f\mu_{def}} \cdot \frac{\rho_{def}}{\rho_{fl}}. \tag{3}$$

For gases at low pressures we can write

$$\frac{\rho_{def}}{\rho_{def}} = \frac{T_{fl}}{T_{def}},$$

whence

$$\mathrm{Re}' = \frac{Gd}{\mu f} \cdot \frac{T_{fl}}{T_{def}}. \tag{4}$$

Consequently, the values of Re calculated by Eqs. (1) and (2) will in general be related by the expression

$$\mathrm{Re}' = \mathrm{Re}'' \frac{\rho_{def}}{\rho_{fl}}, \tag{5}$$

and for gases by

$$\mathrm{Re}' = \mathrm{Re}'' \frac{T_{fl}}{T_{def}}. \tag{6}$$

We will call Re' the velocity value and Re" the mass value. It is apparent from Eq. (6) that when the physical parameters are normalized to the mean flow temperature the values of Re calculated from Eqs. (1) and (2) will be the same.

When the fluid is heated

$$\frac{T_{fl}}{T_{def}} \leqslant 1,$$

hence

$$Re' \leqslant Re'';$$

for cooling of the fluid

$$\frac{T_{fl}}{T_{apr}} \geqslant 1 \quad \text{and} \quad Re' \geqslant Re''.$$

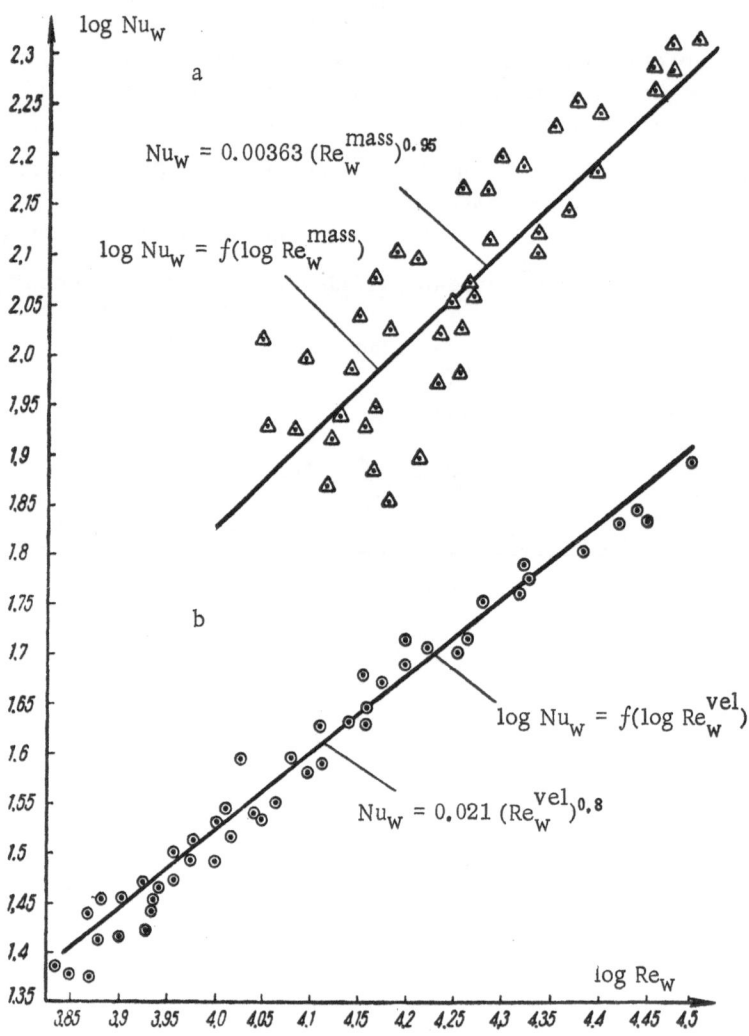

Fig. 1. Results of processing the experimental data of Il'in with normalization of the physical parameters to the wall temperature.

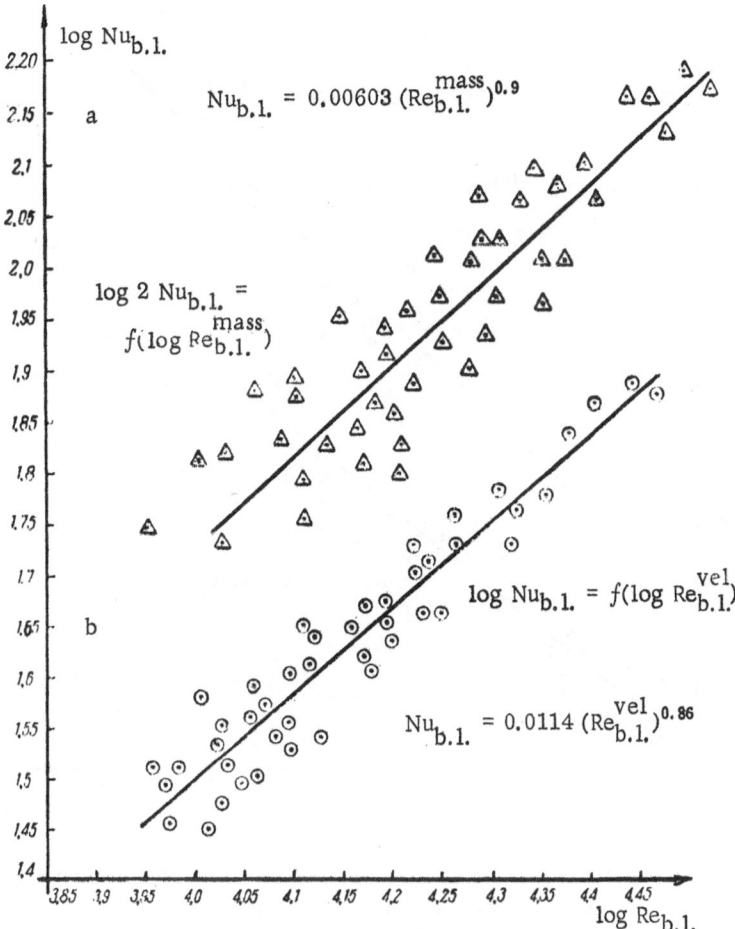

Fig. 2. Results of processing the experimental data of Il'in with normalization of the physical parameters to the boundary layer temperature.

For gases we can write the following expression:

$$Re''^{mass} = \frac{wd}{\nu_{def}} \frac{T_{def}}{T_{fl}} = \frac{w'd}{\nu_{def}},$$

where $w' = wT_{def}/T_{fl}$ is the velocity that the gases would have if the temperature were equal to the defining temperature at the same mass flow and pressure. In the event $T_{def} \neq T_{fl}$ the values of Re' and Re" and, consequently, the invariant heat transfer relations will tend to diverge all the more, the greater the disparity between the temperature simplex T_w/T_{fl} and unity.

We illustrate the above with examples. Figures 1 and 2 give the results of processing the experimental data of Il'in [1] on heat transfer for heated air in a tube 18.31 mm in diameter and with L/D = 65.8. The temperature simplex T_w/T_{fl} varied from 1.081 to 1.84 in the experiments. The Re number was calculated both according to the linear velocity (Figs. 1b and 2b) and according to the mass flow (Figs. 1a and 2a); the physical parameters were normalized to the wall temperature and to the boundary layer temperature.

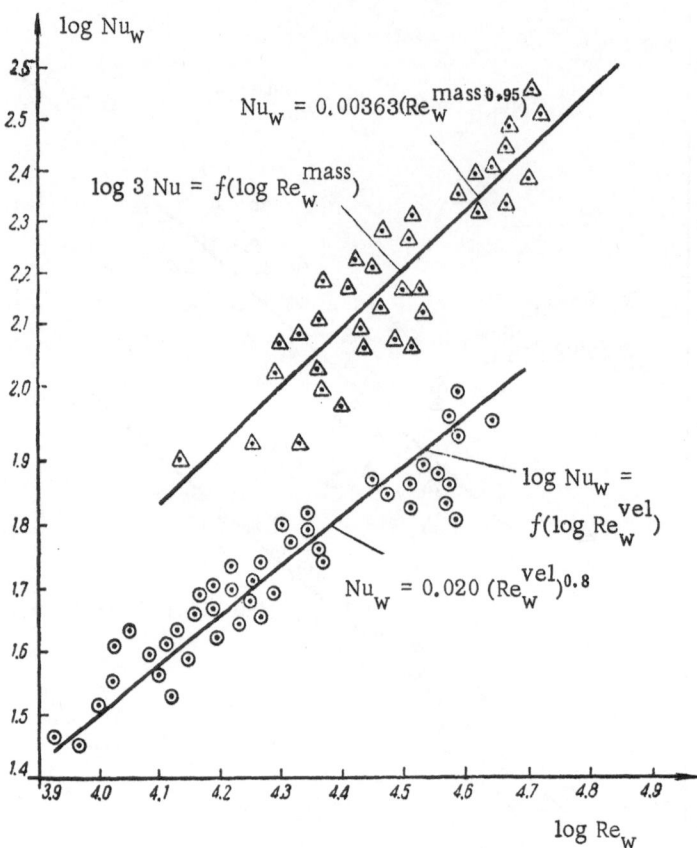

Fig. 3. Results of processing the experimental data of Ivashchenko with normalization of the physical parameters to the wall temperature.

The cited graphical dependences can be approximated by the formulas

$$Nu_W = 0.0210 \, (Re_W^{vel})^{0.8}, \tag{7}$$

$$Nu_{W.} = 0.00363 \, \left(Re_W^{mass}\right)^{0.95}, \tag{8}$$

$$Nu_{b.l.} = 0.0114 \, \left(Re_{b.l.}^{vel}\right)^{0.86}, \tag{9}$$

$$Nu_{b.l.} = 0.00603 \, \left(Re_{b.l.}\right)^{0.91}. \tag{10}$$

Figure 3 shows the results of processing the experimental data of N. I. Ivashchenko [2] on heat transfer in the heating of air in a tube 26.5 mm in diameter and with L/D = 35. The temperature simplex T_W/T_{fl} varied from 1.114 to 2.56. The physical parameters were normalized to the wall temperature. The following relations were discussed:

$$Nu_W = 0.020 \, \left(Re_W^{vel}\right)^{0.8}, \tag{11}$$

$$Nu_W = 0.00363 \, \left(Re_W^{mass}\right)^{0.95}. \tag{12}$$

The invariant relations (7)—(12) turn out to be different when Re is determined in terms of the mass velocity and linear velocity for the same defining temperature. This difference is strongly pronounced when the physical parameters are normalized to the wall temperature, less so when they are normalized to the boundary layer temperature.

If the temperature simplex is contained in the invariant relation, it is then evident from Eq. (4) that the constant coefficient and exponent associated with Re cannot vary in going from the mass to the velocity value of Re. In this case only the exponent of the temperature simplex changes.

The dependence of the viscous friction coefficient on the Re number will also differ depending on whether Re is determined according to Eq. (1) or (2).

In convective heat transfer studies the Reynolds number is calculated both according to Eq. (1) [1, 2, 4, 5] and according to Eq. (2) [3, 6—9] when the physical parameters are normalized to temperatures other than the flow temperature. In [7] justification is given for the preference of Eq. (2) over Eq. (1). However, a detailed analysis of the difference between Eqs. (1) and (2) has not been given in the literature relating to the subject.

Thus, it is indicated in [10] that in processing the experimental data it is necessary in order to minimize the number of computational operations to represent the unknown quantities directly as a function of the measured variables. If the velocity is measured experimentally, then the Re number is determined according to Eq. (1), but if the mass flow is measured the Re number is calculated according to Eq. (2). No mention is made of a difference between the expressions.

In determining the convective component of composite heat transfer by air blowing, Re is calculated in the majority of investigations on the basis of the mass flow, the temperature simplex being left out altogether.

As a result of the foregoing analysis Eq. (1) can be recommended for air blowing, because in this case a smaller scatter of points is obtained when the temperature simplex is significantly different from unity.

In the event that the physical parameters are normalized to a reference temperature other than the wall temperature, the values of the Re number and the invariant relations turn out to be different, depending on whether Re is determined according to the linear velocity or the mass flow rate. A smaller scatter of experimental points is obtained if Re is calculated in terms of the linear velocity.

Literature Cited

1. L. N. Il'in, Effect of Temperature on Convective Heat Transfer, Tr. TsKTI (Transactions of the Central Scientific Research Institute for Boilers and Turbines), No. 4, Book 18 (1950).
2. N. I. Ivashchenko, Effect of the Temperature Factor on Heat Transfer in the Turbulent Flow of a Gas in Tubes, Teploénergetika, No. 2 (1958).
3. N. V. Ilyukhin, Heat Transfer and Friction at High Velocities, Tr. TsKTI, No. 1, Book 2 (1947).
4. V. L. Lel'chuk and B. V. Dedyakin, Heat Transfer from the Wall to a Turbulent Flow of Air in a Tube and Hydraulic Friction at Large Temperature Differences, Coll.: Heat Transfer Problems (Izd. Akad. Nauk SSSR, ÉNIN, 1959).
5. Heat Engineering Handbook, Vol. 1 (Gosénergoizdat, 1957).
6. A. A. Gukhman and N. V. Ilyukhin, Heat Transfer in the Motion of a Gas in Tubes at High Velocities, Tr. TsKTI, Book 12 (1949).
7. A. A. Gukhman et al., Experimental Investigation of Heat Transfer and Friction in the Subsonic Range, Tr. TsKTI, Book 21 (1951).
8. P. K. Konakov, S. S. Filimonov, and B. A. Khrustalev, Heat Transfer in Steam Boiler Combustion Chambers (Rechizdat, 1960).
9. A. V. Arseev, Heat Transfer and Friction in Tubes with High-Temperature Heating of the Air, Teploénergetika, No. 6 (1959).
10. M. A. Mikheev, Fundamentals of Heat Transfer (Gosénergoizdat, 1956).

HEAT TRANSFER BY THERMAL CONDUCTION AND RADIATION
IN AN ABSORBING MEDIUM

Yu. P. Sidorov

Consider heat transfer between two black coaxial cylinders separated by a gray medium with wall temperatures T_1 and T_2.

The energy transport equation is written in the form

$$\frac{d^2\left(T + \frac{\sigma_0}{\kappa\lambda}T^4\right)}{dr^2} + \frac{1}{r}\frac{d\left(T + \frac{\sigma_0}{\kappa\lambda}T^4\right)}{dr} = 0. \tag{1}$$

Solving this equation and determining the arbitrary constants from the boundary conditions implied by the given wall temperatures, we arrive at the following equation:

$$T + \frac{\sigma_0}{\kappa\lambda}T^4 = T_1 + \frac{\sigma_0}{\kappa\lambda}T_1^4 + \frac{(T_2 - T_1) + \frac{\sigma_0}{\kappa\lambda}\left(T_2^4 - T_1^4\right)}{\ln\frac{R_2}{R_1}}\ln\frac{r}{R_1}, \tag{2}$$

or, dividing both sides of the equation by $(T_1^4)\,\sigma_0/k\lambda$, we obtain the following in relative form:

$$Nt + t^4 = 1 + N + \frac{N(t_2 - 1) + \left(t_2^4 - 1\right)}{\ln\frac{R_2}{R_1}}\ln\frac{r}{R_1}, \tag{3}$$

where $N = k\lambda/\sigma_0 T_1^3$ characterizes the ratio of heat transfer due to conduction to the heat transfer due to radiation.

The solution of Eq. (3) is represented in graphical form (Fig. 1). Inspection of the graph reveals that the temperature gradient at the cold wall is always greater than the gradient due to thermal conduction alone and increases as N decreases. Viskanta, Grosh, and Genzel arrive at the same conclusion after analyzing the problem of a plane layer in detail [2, 3].

We have analyzed the problem of the temperature distribution between two black surfaces. However, such surfaces never occur in practice, and it is therefore desirable to investigate the temperature distribution between two gray coaxial cylinders. In this case it is necessary to introduce the concept of an equilibrium layer, inside which the laws of black body radiation will be valid. In the immediate vicinity of the walls temperature continuity can only be maintained by virtue of molecular conduction. Making use of Eq. (1) and assuming its validity for the layers near the wall with temperatures T_{δ_1} and T_{δ_2}, we obtain a solution for the temperature distribution in the middle layer:

$$T + \frac{\delta_0}{\kappa\lambda}T^4 = T_{\delta 1} + \frac{\sigma_0}{\kappa\lambda}T_{\delta 1}^4 - \frac{(T_{\delta 1} - T_{\delta 2}) + \frac{\sigma_0}{\kappa\lambda}\left(T_{\delta 1}^4 - T_{\delta 2}^4\right)}{\ln\frac{r_{\delta 2}}{r_{\delta 1}}}\ln\frac{r}{r_{\delta 1}}. \tag{4}$$

97

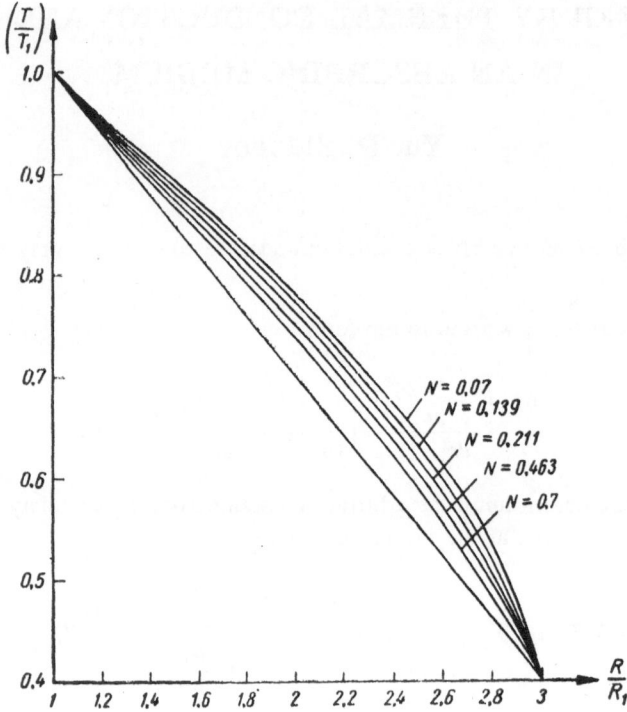

Fig. 1. Temperature distribution in a cylindrical channel
with black walls.

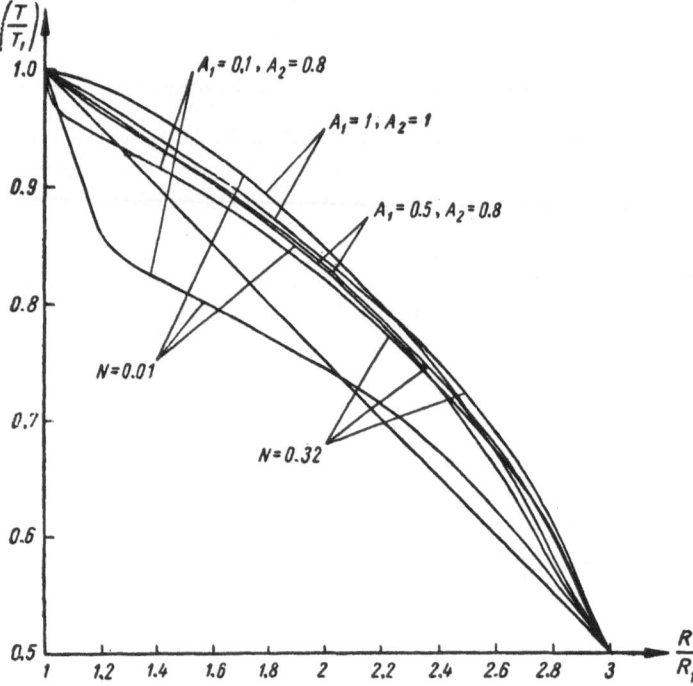

Fig. 2. Temperature distribution between gray walls.

The expression for the heat flux in the middle layer is written as

$$q = \frac{2\pi\lambda\,(T_{\delta 1} - T_{\delta 2})}{\ln\dfrac{r_{\delta 2}}{r_{\delta 1}}} + \frac{2\pi\dfrac{\sigma_0}{\kappa}\left(T_{\delta 1}^4 - T_{\delta 2}^4\right)}{\ln\dfrac{r_{\delta 2}}{r_{\delta 1}}}\,. \tag{5}$$

Analyzing the transfer of energy in the wall layers, the same heat flux may be expressed in the form

$$q = \frac{2\pi\lambda\,(T_1 - T_{\delta 2})}{\ln\dfrac{r_{\delta 1}}{r_1}} + \frac{\sigma_0\left(T_1^4 - T_{\delta 1}^4\right)\,2\pi r_1}{\dfrac{r_1}{r_{\delta 1}}\left(\dfrac{1}{A_1} - 1\right) + \dfrac{1}{2}}\,, \tag{6}$$

$$q = \frac{2\pi\lambda\,(T_{\delta 2} - T_2)}{\ln\dfrac{r_2}{r_{\delta 2}}} + \frac{\sigma_0\left(T_{\delta 2}^4 + T_2^4\right)\,2\pi r_{\delta 2}}{\dfrac{r_{\delta 2}}{r_2}\left(\dfrac{1}{A_2} - 1\right) + \dfrac{1}{2}}\,. \tag{7}$$

Dividing Eqs. (5), (6), and (7) by the quantity$[(T_1^4)\,\sigma_0/k\lambda]$and eliminating the heat flux q, we write the system of equations

$$\left(\frac{N}{\ln\dfrac{r_2 - \dfrac{1}{\kappa}}{r_1 + \dfrac{1}{\kappa}}} + \frac{N}{\ln\dfrac{r_1 + \dfrac{1}{\kappa}}{r_1}}\right)t_{\delta 1} + \left(\frac{1}{\ln\dfrac{r_2 - \dfrac{1}{\kappa}}{r_1 + \dfrac{1}{\kappa}}} + \frac{r_1\kappa}{\dfrac{r_1}{r_{\delta_1}}\left(\dfrac{1}{A_1} - 1\right) + \dfrac{1}{2}}\right)t_{\delta 1}^4 =$$

$$= \ln\frac{N}{\ln\dfrac{r_2 - \dfrac{1}{\kappa}}{r_1 + \dfrac{1}{\kappa}}}t_{\delta 2} + \frac{t_{\delta 2}^4}{\ln\dfrac{r_2 - \dfrac{1}{\kappa}}{r_1 + \dfrac{1}{\kappa}}} + \frac{N}{\ln\dfrac{r_1 + \dfrac{1}{\kappa}}{r_1}} + \frac{r_1\kappa}{\dfrac{r_1}{r_{\delta 1}}\left(\dfrac{1}{A_1} - 1\right) + \dfrac{1}{2}}\,, \tag{8}$$

$$\left(\frac{N}{\ln\dfrac{r_2 - \dfrac{1}{\kappa}}{r_1 + \dfrac{1}{\kappa}}} + \frac{N}{\ln\dfrac{r_2}{r_2 - \dfrac{1}{\kappa}}}\right)t_{\delta 2} + \left(\frac{1}{\ln\dfrac{r_2 - \dfrac{1}{\kappa}}{r_1 + \dfrac{1}{\kappa}}} + \frac{r_{\delta 2}\kappa}{\dfrac{r_{\delta 2}}{r_2}\cdot\left(\dfrac{1}{A_2} - 1\right) + \dfrac{1}{2}}\right)t_{\delta 2}^4 =$$

$$= \frac{N}{\ln\dfrac{r_2 - \dfrac{1}{\kappa}}{r_1 + \dfrac{1}{\kappa}}}t_{\delta 1} + \frac{t_{\delta 1}^4}{\ln\dfrac{r_2 - \dfrac{1}{\kappa}}{r_1 + \dfrac{1}{\kappa}}} + \frac{Nt_2}{\ln\dfrac{r_2}{r_2 - \dfrac{1}{\kappa}}} + \frac{r_{\delta 2}\,\kappa t_2^4}{\dfrac{r_{\delta 2}}{r_2}\left(\dfrac{1}{A_2} - 1\right) + \dfrac{1}{2}}\,. \tag{9}$$

Solution of the system makes it possible to find out the temperatures in the near-wall layers. Then, assuming black body radiation in the middle zone and making use of Eq. (4), we construct the temperature field between the wall layers. Temperature continuity is maintained at the walls by thermal conduction.

Curves showing the temperature distribution between coaxial cylinders with different wall surface emissivities ε are given in Fig. 2.

An analysis of the graph shows that for emissivities not equal to one the molecular temperature near the walls is somewhat lower than in the case of black body radiation. As the wall emissivity is diminished the temperature difference in the wall layer and at the wall becomes more pronounced. This in turn affects the temperature distribution inside the medium.

As the optical density is increased the temperature "jump" is less marked. This permits us to assume black body radiation in an equilibrium layer with high optical density, which means that the temperature curve for black body radiation can be extended to the walls in this case.

Literature Cited

1.   P. K. Konakov, S. S. Filimonov, and B. A. Khrustalev, Heat Transfer in Steam Boiler Combustion Chambers (Rechizdat, 1960).
2.   R. Viskanta and R. I. Gorsh, Heat Transfer by Simultaneous Conduction and Radiation in an Absorbing Medium, Trans. ASME, Series C: J. Heat Transfer, Vol. 83 (1961).
3.   L. Genzel, Der Anteil der Warmestrahlung bei Warmeleistungsvorgänge [Contribution of Thermal Radiation in Heat Transfer Processes], Z. Physik, 135 (177):(1953).

PART II

Equipment for the Preparation of Semiconductor Materials

TEMPERATURE FIELD AND DISLOCATIONS IN A SINGLE CRYSTAL

G. E. Verevochkin

The preparation of single crystals of semiconductor materials with a specified distribution of impurities and a crystal lattice with a high degree of perfection is one of the principal objectives of semiconductor metallurgy. The considerable number of methods for preparing single crystals is patent. At the present time the seed pulling technique (Czochralski method) is used primarily; to a lesser extent, the methods of zone melting and directional crystallization.

The first method has quite a number of advantages, but the single crystals obtained by pulling from the melt are not without defects and are usually nonhomogeneous in their composition. The crystal lattice is always imperfect to a certain extent, containing structural defects [1]. The most important of these from the viewpoint of their influence on the properties of the semiconductor are dislocations. The theory of dislocations is discussed in detail in [2].

Among the myriad factors affecting the onset of defects in the crystal lattice of an ingot, most researchers place considerable stress on the cooling process of the crystal and the formation of the temperature field in the latter.

Petrov [3] states that the reason for the appearance of dislocations in single-crystal ingots is to be found in thermal stresses, which occur as the result of nonuniform cooling. The author feels that the choice of thermal conditions and manner in which they are maintained during growth and subsequent cooling of the ingot must be uppermost in the technology of growing single crystals.

The growth of single crystals from the molten phase, remarks Tannenbaum [4], implies, in principle, a solution of the problem of controlling the temperature and temperature gradients in the melt and in the ingot.

Unfortunately, insufficient theoretical and experimental research has been accomplished to date in this area. Research efforts are directed by and large toward the design of reliable semiconductor devices based on perfect single crystals.

Billig [5] considers that thermal stresses in the ingot are the reason for the onset of dislocations in crystals prepared by seed pulling. He has analyzed the problem and suggests the following formula as a quantitative estimate of the dislocation density in a pulled single crystal with temperature gradients present:

$$n = \frac{\alpha}{b} \frac{\delta T}{\delta r} , \qquad (1)$$

where n is the dislocation density, $\delta T / \delta r$ is the radial temperature gradient, b is the Burgers vector.

The above formula was qualitatively confirmed in experiments with silicon and germanium ingots. Billig discovered that the dislocation density is a maximum at the surface of the ingot, the central protion being relatively free of defects, i.e., the number of etch pits is found to depend directly on the quantity $\delta T / \delta r$.

Indenbom [6] performed a theoretical analysis of the onset of dislocations in crystal growth. The author also feels that these defects are a consequence of nonuniformity in the temperature distribution within the ingot. Contrary to Billig, he supposes that the main portion of the dislocations arise in a growing single crystal due to axial, rather than radial, temperature gradients, since axial gradients are the most important in the pulling process.

Indenbom has shown that the dislocation density is a tensor quantity

$$\hat{\beta} = \hat{\alpha} \, \text{grad} \, T, \tag{2}$$

where $\hat{\beta}$ is the dislocation density tensor, $\hat{\alpha}$ is the tensor coefficient of thermal expansion of the crystal.

For a material with isotropic α the tensor $\hat{\beta}$ has the following form in cylindrical coordinates θ, r, z:

$$\alpha \left\{ \begin{matrix} 0 & \dfrac{\partial}{\partial z} & 0 \\[2mm] -\dfrac{\partial}{\partial z} & 0 & \dfrac{\partial}{\partial r} \\[2mm] 0 & -\dfrac{\partial}{\partial r} & 0 \end{matrix} \right\} T(r, z). \tag{3}$$

The coefficients of the tensor $\hat{\beta}$ are equal, respectively, to

$$\begin{aligned} \beta_{rr} &= \beta_{\theta\theta} = \beta_{zz} = 0, \\[1mm] \beta_{r\theta} &= -\beta_{\theta r} = \alpha \frac{\partial T}{\partial z}, \\[1mm] \beta_{\theta z} &= -\beta_{z\theta} = \alpha \frac{\partial T}{\partial r}. \end{aligned} \tag{4}$$

The experimental work of Penning [7] is devoted to a qualitative investigation of the effects of temperature gradients on the onset of dislocations in single crystals. He performed experiments with cylindrical single crystals of germanium approximately 20 mm in diameter.

By uniform cooling of the surface of the cylindrical ingot a radial temperature gradient was created, a temperature gradient along the axis being produced only by the emission of heat from the end surfaces of the sample.

In the first series of experiments the single crystal was heated to a temperature of 850°C, then it was immersed in a container of liquid tin at a temperature of 400°C. The thermal stresses were so large that surface cracks appeared on the samples. The largest number of such defects were observed on the surface of the crystal, decreasing rapidly toward the center. However, the number of dislocations did not increase appreciably.

In the second series of experiments the cooling rate was made much lower. The single crystals were cooled in a slow stream of hydrogen.

In the investigation of the samples, defects in the form of surface cracks were not disclosed, but the number of etch pits increased considerably, appearing in straight lines of different densities with respect to the crystal cross section. The region of minimum etch pit density lay between the center and surface of the sample and was annular in shape.

The effect of the nonuniform axial temperature gradient on the dislocation density was investigated in the next experiment. A germanium single crystal was prepared by growing from the melt by the Czochralski method. After it had attained several centimeters in length the crystal was removed from the melt and conditions were established whereby the required temperature difference was obtained along the axis of the cylinder. Then the sample was cooled to room temperature in a quiescent inert gas atmosphere.

After etching of a transverse section of the single crystal a considerable increase in the dislocation density was observed. The region of low etch pit density was triangular.

Penning thus showed experimentally that imperfections of germanium single crystals in the form of dislocations can appear as the result of both radial and axial temperature gradients.

The effects elicited by temperature gradients in the pulled single crystal cannot be oversimplified as mere cooling of the ingots used in the work just described. It is important, however, that Penning set forth the possibility of forming defects as the result of nonuniform cooling of the ingots independently of the direction of heat emission.

Appropriate to this is a critical reference to the conclusions in [5], where only the radial temperature gradient is included in the calculation of the dislocation density. It is known from the theory of elasticity [8] that when a temperature gradient is present in a body, thermal stresses will be absent therein only if the second derivative of the temperature with respect to a given direction is equal to zero, i.e., when the temperature depends linearly on the distance. As will be shown, in single crystals pulled from the melt the variation in temperature along the axis of the ingot does not obey a linear law. Consequently, for a more rigorous accounting of the factors promoting the formation of defects in a crystal it is necessary to include both components $\partial T/\partial r$ and $\partial T/\partial z$, as Indenbom has done.

The influence of the cooling conditions on the formation of dislocations in germanium single crystals has been investigated by Kokorish [9–11].

Germanium ingots were cooled very rapidly during their growth from the melt, by lowering a special running-water cooling element into the pulling furnace. The inside diameter of the element was 25 mm. The pulling was accomplished in an atmosphere of hydrogen at a velocity of 1.5 mm/min.

Data are given in [9] on the dislocation density in crystals prepared by pulling under ordinary conditions and with the cooling element present. All of the single crystals grown under rapid cooling conditions had a dislocation density several orders of magnitude higher than the ingots prepared under normal circumstances. In the experiments using the cooling element a sharp rise in the etch pit density was observed at the surface, an effect which did not appear in the crystals pulled under normal conditions.

E. Yu. Kokorish asserts that the onset of dislocations is affected by radial as well as axial temperature gradients. He attributes the abrupt rise in surface etch pit density on the crystal in particular to the presence of a large radial gradient; the increased dislocation density at the beginning of the ingot is ascribed to the temperature gradient along the crystal axis.

In [10] investigations were performed on the emission of heat from the crystallization front by means of the Peltier effect. Germanium single crystals were prepared by pulling from the melt in an atmosphere of hydrogen at a velocity of 1 mm/min; the rate of rotation of the seed crystal was 60 rpm, the current density 23 A/cm^2.

The investigations were carried out by a comparison of the ingots prepared under three different sets of conditions: a) current off (ordinary conditions); b) current in forward direction (plus on crystal); c) current in reverse direction (minus on crystals).

All of the crystals prepared under ordinary conditions had the highest dislocation density. Fewer defects occurred in the ingots in the forward current experiments. Also, under these conditions of growth a more uniform distribution of the dislocations was noted, both in length and in cross section of the sample.

The influence of growth rate, crystal diameter, quality of the seed, and cooling conditions on the structural perfection of germanium single crystals prepared by seed pulling was investigated in [11]. The authors of that paper postulate that the conditions under which the ingot is cooled from the hardening temperature to room temperature play the fundamental role in the onset of dislocations in germanium single crystals.

Experiments have been conducted on the growth of germanium crystals from the melt in a special crucible, the walls of which doubled as a heat shield; the aim of the experiments was to eliminate large temperature gradients. The axial gradient on the surface of the crystal was 20–40 deg/cm. The dislocation density in the samples prepared under such conditions did not exceed 10 cm^{-2} in the predominant direction.

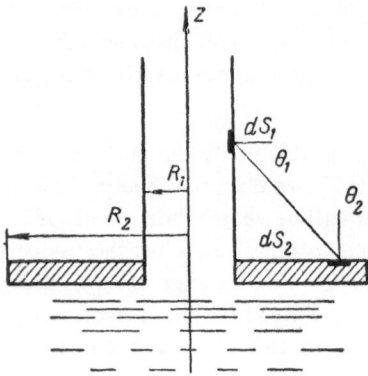

Fig. 1.   Shielding arrangement.

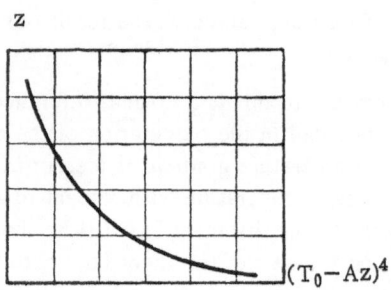

Fig. 2.   Graph of the dependence
$(T-Az)^4 = \varphi(z)$.

In an experimental study Rosi [12] also focused attention on dislocations and their dependence on the temperature gradients. The experimental melting was carried out in an atmosphere of pure helium, the pulling rate was 1 mm/min, the rate of seed rotation 25 rpm.

Single-crystal ingots were prepared with temperature gradients from 10 to 220 deg/cm (see [12], Table 1). The gradients were measured with a thermocouple situated at the center of the test specimen at a distance of 1 cm from the phase interface.

Interesting investigations have been conducted by Francois [13]. For the elimination of the thermal stresses occurring in a crystal during its cooling, it is necessary and sufficient for the crystal temperature T to be independent of the radius and angle φ, i.e., for the isotherms to be plane, and the value of T to vary in the direction of the crystal axis according to a linear law:

$$T = T_0 - Az,$$

where T_0 is the temperature of crystallization of the material, A is the temperature gradient in the crystal.

The necessary conditions on the surface are contained in the following:

1. The radial thermal flux must be equal to zero:

$$q = -\lambda \frac{\partial T}{\partial r} = 0. \tag{5}$$

An elementary surface area of the crystal must give off and receive the same amount of heat from a nearby source.

2. The temperature on the surface of the ingot must decrease linearly with height:

$$T = T_0 - Az.$$

The author investigates the elementary heat balance (Fig. 1).

The quantity of heat emitted by an area dS_1 into a half-space is

$$dQ = \varepsilon\sigma T_1^4 ds_1, \tag{6}$$

where ε_1 is the emissivity of the crystal, σ is the black body (Stefan) radiation constant, T_1 is the temperature of the given point on the crystal, dS_1 is an element of area on the crystal at that point.

The quantity of heat received by dS_1 from the element dS_2 is

$$\iint a_1\varepsilon_2 \frac{\sigma}{\pi} T_2^4 ds_1 ds_2 \frac{\cos\theta_1\cos\theta_2}{p^2}, \tag{7}$$

where a_1 is the radiant energy absorption coefficient, ε_2 is the surface emissivity, T_2 is the surface temperature, p is the distance between the surfaces, dS_2 is the emitting surface.

It is assumed that the quantity T_2 is known and constant over the entire surface. Then the integral $\iint \frac{\cos\theta_1 \cos\theta_2}{p^2} dS_2$ is determined by the geometry of the system. Invoking the axial symmetry of the prob-

lem, denoting $\iint \frac{\cos\theta \cos\theta_2}{p^2} ds_2 = F(z)$, and assuming that $a_1 = \varepsilon_1$, the author derives the fundamental equation

$$\pi (T_0 - Az)^4 = \varepsilon_2 F(z) T_2^4. \tag{8}$$

The rest of the investigation reduces to a calculation of the values of the function F(z) for various surfaces of cylindrical symmetry.

The effectiveness of each surface is evaluated by comparison of the function F(z) with $(T_0 - Az)^4$, a graph of which is shown in Fig. 2 as a function of the coordinate z. It is necessary in order for the condition (8) to be satisfied that these functions coincide.

The most effective surface turns out to be the one shown in Fig. 1.

The function F(z) in this case has the form

$$F(z) = \text{arc tg} \frac{\sqrt{R_2^2 - R_1^2}}{z} + \frac{z}{2R_1} \text{ arc cos } \frac{R_1}{R_2} -$$

$$- \frac{z(R_2^2 - R_1^2 + z^2)}{R_1} \frac{1}{X} \text{arc tg} \left[\frac{\sqrt{R_2^2 - R_1^2}}{R_2 + R_1} \right] \frac{x}{y}, \tag{9}$$

where x = $[(R_2 + R_1)^2 + z^2]^{1/2}$; y = $[(R_2 - R_1)^2 + z^2]^{1/2}$; R_1 is the radius of the crystal; and R_2 is the largest radius of the ring.

The surface temperature of the ring, equal to

$$T_2 = 1.19 T_0,$$

is found from the condition $z \to 0$ on the assumption that $\varepsilon_2 = 1$.

The analytical part was tested experimentally in the growing of germanium single crystals by the Czochralski method. A graphite ring with a thickness $\delta = 1.0$ mm with an outside diameter equal to 48 mm and an inside diameter of 20 mm was used in the experiments. The difference between the crystal diameter and ring was 1~2 mm. Using the ring, the dislocation density was diminished by approximately one order of magnitude. Francois notes in conclusion that reproduction of the crystals is very difficult to obtain.

It is apparent from the investigations discussed that the formation of single crystals is strongly influenced by the heat transfer conditions and temperature field in the ingot.

Researchers have failed, however, to be adequately concerned with this problem. The literature to date contains no published experimental works devoted to the systematic investigation of the temperature field in single crystals. The theoretical solutions discussed below are approximate and cannot be accepted as conclusive, due to the complexity of the boundary conditions and the lack of precise values for the theoretical variables and parameters involved (thermal conductivity, emissivity of the material, etc.).

The temperature distribution in a circular cylinder of radius r_0 and semi-infinite length in the z direction is analyzed in [5]. One end of the cylinder is maintained at a constant temperature T_0 by molten metal. The heat conduction equation in cylindrical coordinates under steady-state conditions is written in the form

$$\frac{\partial^2 T}{\partial r^2} + \frac{1}{r} \cdot \frac{\partial T}{\partial r} + \frac{\partial^2 z}{\partial z^2} = 0. \tag{10}$$

A plane crystallization front is assumed at the phase interface, as reflected in the boundary conditions

$$T = T_0 \quad \text{for} \quad z = 0 \quad \text{and} \quad r_0 \geqslant r \geqslant 0. \tag{11}$$

The heat transfer conditions at the crystal surface are defined by the law

$$\kappa \frac{\partial T}{\partial r} = HT \quad \text{for} \quad r = r_0 \quad \text{and} \quad 0 < z < \infty, \tag{12}$$

where T_0 is the crystallization temperature of the material, k is the thermal conductivity, H is the emissive power of the material.

To find the solution it was assumed that

$$h = \frac{H}{\kappa} \approx \text{const.} \tag{13}$$

The analytical solution of Eq. (10) was obtained in the form of an infinite series of first- and second-order Bessel functions:

$$\frac{T}{T_0} = 2r_0 h \sum_{n=1}^{\infty} \frac{I_0(\alpha_n r)}{I_0(\alpha_n r)} \cdot \frac{e^{-\alpha_n z}}{(r_0 h)^2 + (r_0 \alpha_n)^2}. \tag{14}$$

For a rough estimate of the temperature distribution in a crystal the following equation is proposed:

$$\frac{T}{T_0} \approx \left[1 - \frac{r_0 h}{2} \left(\frac{r}{r_0} \right)^2 \right] \exp \left[-\frac{z}{r_0} (2r_0 h)^{1/2} \right]. \tag{15}$$

Equation (15) implies, as a first approximation, a parabolic temperature distribution with respect to the radius of the ingot. The radial nonuniformity is greater, the higher the values of h and r. In the direction of the crystal axis the temperature decreases exponentially. The temperature distribution along the radius and axis of the crystal is calculated according to Eq. (14) for a silicon ingot with a diameter d = 20 mm.

Boikov and Kuchin [14] describe some of the shortcomings of the work just described. They regard it as an extremely crude assumption that the cooling of the seed crystal is not taken into account in [5]. The authors propose a more precise solution to the heat conduction equation (10). With the boundary conditions

$$T = T_f \quad \text{for} \quad z = l \quad 0 < r < R,$$
$$-\lambda \left(\frac{\partial T}{\partial r} \right)_{r=R} = \alpha (T_R - T_f), \quad 0 < z < l,$$
$$T = T_0 \quad \text{for} \quad z = 0.$$

The solution has the form

$$T(r, z) = T_f + (T_f - T_0) \Sigma A_n I_0 \left(\mu_n \frac{r}{R} \right) \exp \left(-\mu_n \frac{z}{R} \right) \frac{1 - \exp \left(-2\mu_n \frac{l-z}{R} \right)}{1 - \exp \left(-2\mu_n \frac{l}{R} \right)}, \tag{16}$$

where T_0 is the temperature of the surrounding medium, T_f is the temperature of the melt, λ is the thermal conductivity, α is the heat transfer coefficient.

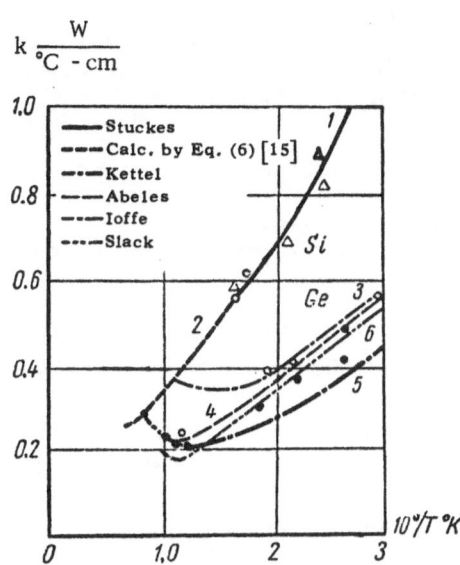

Fig. 3. Graph of the thermal conductivity of silicon and germanium as a function of temperature.

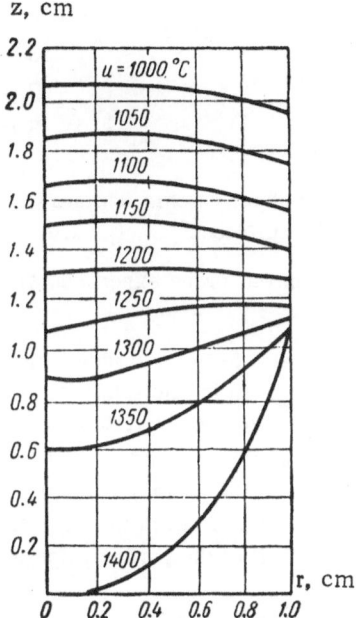

Fig. 4. Temperature distribution in a silicon ingot.

Equation (16) is recommended for approximate practical calculations of the temperature field in a single crystal grown with rapid cooling of the seed.

The differential heat conduction equation for a cylindrical silicon ingot of semi-infinite length in the z direction and radius a is

$$\kappa\left[\frac{\partial^2 u}{\partial r^2} + \frac{1}{r}\frac{\partial u}{\partial r} + \frac{\partial^2 u}{\partial z^2}\right] + \left[\frac{\partial \kappa}{\partial r}\frac{\partial u}{\partial r} + \frac{\partial \kappa}{\partial z}\frac{\partial u}{\partial z}\right] + Q = 0, \tag{17}$$

where u is the temperature of the ingot, k is the coefficient of thermal conductivity of the material, Q is the heat generated in the ingot from high-frequency heating; Akiyama and Yamaguchi [15] solved the equation numerically on an IBM-650 computer for steady-state conditions, using the method of Liebmanus.

The second term of Eq. (17) reflects the dependence of k on the temperature of the ingot. The heat transfer conditions at the surface of the ingot for r = a are described by the boundary conditions

$$\left[\frac{\partial u}{\partial r} + \frac{\varepsilon\sigma}{\kappa}(u+300)^4 + \frac{H}{\kappa}u\right]_{r=a} = 0 \quad \text{for} \quad z > l.$$

The conditions at the boundary z = l and z = 0 have the form

$$u_{z=l} = 0, \quad \text{for} \quad z = l,$$
$$u_{z=0} = \varphi_1(r) \quad \text{for} \quad z = 0, \quad 0 < r < a,$$
$$u_{z=0} = \varphi_2(z) \quad \text{for} \quad r = 0, \quad 0 < z < l_1,$$

where $z = l_1$ is the upper end of the molten zone.

In view of the lack of data in the literature on the thermal conductivity of silicon at high temperatures, the authors of [15] use extrapolated data (Fig. 3).

For the solution of Eq. (17) the entire ingot was divided at 400 points in the r and z directions. The division was made at

$$0 < r < a, \qquad \Delta r = 0.1,$$
$$0 < z < 2.0, \qquad \Delta z = 0.1,$$
$$2.0 < z < 5.0, \quad \Delta z = 0.3,$$
$$5.0 < z < 15.0, \ \Delta z = 15.0.$$

The results of the calculations are shown in Fig. 4 for values of $\alpha = 1.0$ cm, $l = 15.0$ cm, $\varepsilon\sigma = 0.96 \cdot 10^{-12}$ cal $\cdot$ cm^{-2}sec^{-1}°C^4, $l_1 = 1.0$ cm, H = 10^{-8} cm^{-1}.

The solution was tested experimentally by melting small bits of various elements placed in a recess in a silicon crystal.

In the next article Uglov discusses the theoretical solution of the stated problem; an approximate numerical calculation is given for the temperature field in a germanium single crystal. One-dimensional solutions are given in [16-19] for steady-state conditions.

In conclusion, the papers discussed herein demonstrate conclusively the exceptionally important influence of thermal factors on the forming of single crystals. However, so far there have been no reliable analytical relations for the temperature field in single crystals prepared under realistic conditions. Such relations would be useful for the analysis of the various imperfections that occur in the growth of ingots.

Literature Cited

1. B. A. Krasyuk and A. I. Gribov, Germanium and Silicon Semiconductors (Metallurgizdat, 1961).
2. W. T. Read, Dislocations in Crystals [Russian translation] (IL, 1957).
3. D. A. Petrov, Structural Defects in Single Crystal Semiconductors, In: Problems of Metallurgy and Semiconductor Physics (Izd. Akad. Nauk SSSR, 1959).
4. N. B. Hannay, Semiconductors [Russian translation] (IL, 1962).
5. E. Billig, Proc. Roy. Soc., Series A, 235:37- 55 (1956).
6. V. L. Indenbom, The Macroscopic Theory of the Formation of Dislocations in Crystal Growth, Kristallografiya, (2):594-603 (1957).
7. P. Penning, Philips Res. Rep., 13:69 (1958).
8. S. P. Timoshenko, Theory of Elasticity (GTTI, Leningrad-Moscow, 1934).
9. E. Yu. Kokorish, Influence of Cooling Conditions on the Formation of Dislocations in Germanium Crystals, In: Crystal Growth, Vol. 2 (Izd. Akad. Nauk SSSR, 1959).
10. E. Yu. Kokorish, Influence of the Peltier Effect on the Perfection of Germanium Single Crystals Grown by the Method of Pulling from the Melt, Kristallografiya, 5 (5):(1960).
11. E. Yu. Kokorish and N. N. Sheftal', Growth of Dislocation-Free Germanium Single Crystals, In: Crystal Growth, Vol. 3 (Izd. Akad. Nauk SSSR, 1960).
12. F. D. Rosi, RCA Review, 19 (3):349-367 (1958).
13. M. Francois, Solid State Physics in Electronics and Telecommunications, Vol. 1: Semiconductors, p. 171 (1960).
14. G. P. Boikov and V. D. Kuchin, Temperature Field in a Synthetically Grown Single Crystal, Vyssh. Uchebn. Zav. Fiz., No. 2 (1959).
15. K. Akiiama and I. Yamaguchi, J. Appl. Phys., 33 (5):1899-1900 (1962).
16. E. Billig, Proc. Roy. Soc., Series A, 229:346-363 (1955).
17. N. N. Sheftal' (editor), Natural and Synthetic Crystal Growth Processes [collected Russian translations] (IL, 1963).
18. B. M. Gol'tsman, Cooling of Sheets, Tubes, and Thin Rods Pulled from the Melt, Izv. Vyssh. Uchebn. Zav. Fiz., No. 6 (1958).
19. Yu. Shashkov, Metallurgy of Semiconductors (Metallurgizdat, 1960).
20. Technology of Semiconductor Materials, Vol. 1 (Oborongiz, 1961).
21. D. C. Bennet and B. Sawyer, Bell System Tech. J., 35:637 (1956).
22. T. G. Cressel and I. A. Powell, Progress in Semiconductors, Vol. 2 (1957).

TEMPERATURE FIELD IN GERMANIUM SINGLE CRYSTALS
PREPARED BY THE CZOCHRALSKI METHOD

A. A. Uglov

The structural perfection and distribution of impurities in the length and cross section of single crystals pulled from the melt by the Czochralski method are determined largely by the temperature fields of the melt and ingot during the growth process [1, 2]. The thermal process involved in the production of single-crystal germanium is described by a system of differential equations in partial derivatives for the temperature fields of the melt and ingot, which are joined at the phase interface by suitable uniqueness criteria [3]. For simplification of the solution this problem can be broken down into two parts, treating separately the temperature field of the melt and the variation in temperature field of the ingot during the growth process.

In the present article we consider the temperature field of a germanium single crystal in various growth situations. The nonstationary temperature field of an ingot with a radius r_0, pulled at a constant velocity v, is described by the differential heat conduction equation

$$\frac{1}{a} \cdot \frac{\partial t}{\partial \tau} = \frac{\partial^2 t}{\partial r^2} + \frac{1}{r} \cdot \frac{\partial t}{\partial r} + \frac{\partial^2 t}{\partial z^2} \tag{1}$$

for $\tau > 0$, $r_0 > r > 0$, $v\tau \geq z > 0$ in a cylindrical coordinate system with origin at the seed crystal.

At the interface between the solid and liquid phases the temperature remains constant and equal to the crystallization temperature t_0:

$$t = t_0, \quad z = v\tau. \tag{2}$$

Another boundary condition describes the heat exchange between the surface of the single crystal and the surrounding water-jacketed walls of the apparatus, the temperature of which is equal to t_w:

$$-\lambda \frac{\partial t}{\partial r} = \alpha(t - t_w), r = r_0, \tag{3}$$

where λ is the thermal conductivity, α is the heat transfer coefficient.

Since the condition (2) is already specified at the moving boundary, by the substitution of variables $z + z' = v\tau$, $r = r'$, $\tau = \tau'$ we go over to a new coordinate system whose origin is fixed on the phase interface. The initial condition is set forth as

$$t = t_0 \exp(-\kappa z'), \quad \tau' = 0, \tag{4}$$

where the parameter $k \geq 0$ defines the initial temperature distribution. In the simplest case $k = 0$. By the substitution $t = t_0 T$, $r' = r_0 \rho$, $z' = r_0 \xi$, $t_w = t_0 T_w$, $\tau' = Fo(r_0^2/a)$, we bring Eq. (1) and the boundary conditions (2), (3), (4) into dimensionless form.

To solve the system we use the method of integral transforms. Applying the Hankel transform in ρ and the Laplace transform in Fo to the equation and boundary conditions, we readily obtain a second-order ordinary differential equation. Solving the equation and requiring that the solution be bounded as $\xi \to \infty$, we obtain an expression, the inverse transform of which is found by means of tables of inverse transforms [4] and the inverse Hankel transform [5]:

$$\theta\left(\rho,\ \xi,\ \mathrm{Fo}\right)=\mathrm{Bi}\sum_{n=1}^{\infty}\frac{I_0\left(s_n\rho\right)}{I_0\left(S_n\right)\left(S_n^2+\mathrm{Bi}^2\right)}\left\{(1-T_\mathrm{w})\times\right.$$

$$\times\exp\left(\frac{1}{2}\,\mathrm{Pe}\cdot\xi\right)\left[\exp\left(-\xi\,\sqrt{S_n^2+\frac{1}{4}\,\mathrm{Pe}^2}\right)\mathrm{erfc}\left(\frac{\xi}{2\,\sqrt{\mathrm{Fo}}}-\right.\right.$$

$$-\sqrt{\left(S_n^2+\frac{1}{4}\,\mathrm{Pe}^2\right)\mathrm{Fo}}\,\Big)+\exp\left(\xi\,\sqrt{S_n^2+\frac{1}{4}\,\mathrm{Pe}^2}\right)\mathrm{erfc}\left(\frac{\xi}{2\,\sqrt{\mathrm{Fo}}}+\right.$$

$$+\sqrt{\left(S_n^2+\frac{1}{4}\,\mathrm{Pe}^2\right)\mathrm{Fo}}\,\Big)\Big]+T_\mathrm{w}\cdot\exp\left(\frac{1}{2}\,\mathrm{Pe}\cdot\xi-S_n^2\cdot\mathrm{Fo}\right)\times$$

$$\times\left[\exp\left(\frac{1}{2}\,\mathrm{Pe}\cdot\xi\right)\mathrm{erfc}\left(\frac{\xi}{2\,\sqrt{\mathrm{Fo}}}+\frac{1}{2}\,\mathrm{Pe}\,\sqrt{\mathrm{Fo}}\right)+\right.$$

$$+\exp\left(-\frac{1}{2}\,\mathrm{Pe}\cdot\xi\right)\mathrm{erfc}\left(\frac{\xi}{2\,\sqrt{\mathrm{Fo}}}-\frac{1}{2}\,\mathrm{Pe}\,\sqrt{\mathrm{Fo}}\right)\Big]-\exp\left[\frac{1}{2}\,\mathrm{Pe}\cdot\xi-\right.\tag{5}$$

$$-\mathrm{Fo}\left(S_n^2-\mathrm{Pe}\cdot\varkappa r_0-\varkappa^2 r_0^2\right)\Big]\times\left\{\exp\left[-\xi\left(\varkappa r_0+\frac{1}{2}\,\mathrm{Pe}\right)\right]\times\right.$$

$$\times\mathrm{erfc}\left(\frac{\xi}{2\,\sqrt{\mathrm{Fo}}}-\left(\varkappa r_0+\frac{1}{2}\,\mathrm{Pe}\right)\sqrt{\mathrm{Fo}}\right)+\exp\left[\xi\left(\varkappa r_0+\right.\right.$$

$$+\frac{1}{2}\,\mathrm{Pe}\right)\Big]\mathrm{erfc}\left[\frac{\xi}{2\,\sqrt{\mathrm{Fo}}}+\left(\varkappa r_0+\frac{1}{2}\,\mathrm{Pe}\,\sqrt{\mathrm{Fo}}\right)\right]-2T_\mathrm{w}\cdot\exp\left(-S_n^2\mathrm{Fo}\right)+$$

$$+2\exp\left[-\varkappa r_0\xi-\mathrm{Fo}\left(S_n^2-\mathrm{Pe}\cdot\varkappa r_0-\varkappa^2 r_0^2\right)\right]\Big\}\Big\},$$

where $\theta=(t-t_\mathrm{w})/t_0$, $\mathrm{Bi}=\alpha r_0/\lambda$ is the Biot number, $\mathrm{Fo}=a\tau/r_0^2$ is the Fourier number, $\mathrm{Pe}=vr_0/a$ is the Péclet number, and S_n are the roots of the equation

$$SI_1\left(S\right)-\mathrm{Bi}I_0\left(S\right)=0.\tag{6}$$

The nonstationary temperature field of the ingot is a complex function of the numbers Fo, Bi, Pe, and the relative variables ξ and ρ. Equation (5) is simplified considerably for large values of Fo. This has to do with the fact that the functions erfc z converge rather rapidly to their limiting values for large z. It can be shown that under these conditions regular thermal behavior will be observed in the growing of germanium single crystals. In the series sum (5) the root S_1 of Eq. (6) is much smaller than the second and following root. Consequently, in the series sum it is sufficient to retain just the first term and ignore all the rest. Numerical estimates show that the error incurred thereby does not exceed a few tenths of a percent. Then the time variation in the temperature field of the single crystal will be described by the simple exponential relation

$$\theta=2\mathrm{Bi}\left(1-T_\mathrm{w}\right)\exp\left[\left(\frac{1}{2}\,\mathrm{Pe}-\sqrt{S_1^2+\frac{1}{4}\,\mathrm{Pe}^2}\right)\right]\xi\,\frac{I_0\left(S_1\rho\right)}{I_0\left(S_1\right)\left(S_1^2+\mathrm{Bi}^2\right)}.\tag{7}$$

In real crystal-growing situations the heat transfer coefficient α is a complex function of the temperature. If we assume that the emission from the surface of the crystal follows the Stefan–Boltzmann law, the heat transfer coefficient is a function of the temperature cubed:

$$\alpha=\sigma_0\varepsilon_\mathrm{N}\tilde\gamma\left(t,\ t_\mathrm{w}\right)\varphi_{12},\tag{8}$$

where $\varepsilon_\mathrm{N}=\dfrac{1}{1+\varphi_{12}\left(\dfrac{1}{\varepsilon}-1\right)+\varphi_{21}\left(\dfrac{1}{\varepsilon_\mathrm{w}}-1\right)}$ is the normalized emissivity, ε is the emissivity of germanium, ε_w

is the surface emissivity of the walls, φ_{12} and φ_{21} are angular (directional) coefficients.

The quantity $\gamma(t, t_W)$ is defined as

$$\gamma(t,\ t_W) = t^3 + t^2 t_W + t t_W^2 + t_W^3.$$

Equation (8) can be used to calculate the temperature field of an ingot either under conditions of partial shielding or without shielding. If there is no shielding of the crystal, the expression for the heat transfer coefficient has a simpler form, because then

$$\varphi_{12} \simeq 1, \text{ and } \varphi_{21} \ll 1, \quad \alpha = \sigma_0 \varepsilon \gamma(t,\ t_W). \tag{9}$$

Shielding of the grown ingot is used in order to reduce the radial and axial temperature gradients, since large gradients lead to sizable thermal stresses in the ingot. Thermal stresses deteriorate the structural perfection of the single crystal, causing the onset of dislocations (as many as 10^4 cm^{-2}). Knowledge of the temperature fields of single crystals under various shielding arrangements should help to find the optimum conditions under which the number of structural imperfections are minimized.

Under conditions of shielding the exact solution of the problem of the temperature field in an ingot will differ from Eq. (5). However, it would be extremely difficult to find its exact analytical form. For this reason numerical methods are used in [6] to calculate the temperature field of a silicon ingot. Numerical calculations are also seriously impeded by the lack of experimental data on the thermophysical constants of germanium (thermal conductivity, specific heat, emissivity) near the crystallization point.

Numerical calculations are made using a procedure that takes account of the temperature dependence of the dimensionless criteria characterizing the thermal process. We assume that with the use of shields the form of the function (5) remains unchanged and that a change in the growth conditions is manifested only in the value of the heat transfer coefficient. We divide the crystal into short intervals, treating the thermophysical constants and heat transfer coefficient in each interval as independent of the temperature. Then Eq. (5) will be valid for each interval and is then used for the calculations (with k = 0):

$$\frac{t - t_W}{t_0 - t_W} = \text{Bi}\,\frac{I_0(S_1\rho)}{I_0(S_1)(S_1^2 + \text{Bi}^2)} \times \left\{ \exp\left[(\text{Pe} - \mu_1)\left(\text{PeFo} - \frac{z}{r_0}\right)\right] \times \right.$$
$$\times \text{erfc}\left[(\text{Pe} - \mu_1)\sqrt{\text{Fo}} - \frac{z}{2r_0\sqrt{\text{Fo}}}\right] + \exp\left[\mu_1\left(\text{PeFo} - \frac{z}{r_0}\right)\right] \times$$
$$\times \text{erfc}\left(\mu_1\sqrt{\text{Fo}} - \frac{z}{2r_0\sqrt{\text{Fo}}}\right) - \exp\left[(\text{Pe}^2 - S_1^2)\text{Fo} - \frac{z\text{Pe}}{r_0}\right] \times$$
$$\times \text{erfc}\left(\text{Pe}\sqrt{\text{Fo}} - \frac{z}{2r_0\sqrt{\text{Fo}}}\right) + \exp(-S_1^2\text{Fo}) \times$$
$$\left. \times \left[2 - \text{erfc}\left(-\frac{z}{2r_0\sqrt{\text{Fo}}}\right)\right]\right\}, \tag{10}$$

where

$$\mu_1 = \frac{\text{Pe}}{2} + \sqrt{S_1^2 + \frac{\text{Pe}^2}{4}}.$$

The initial temperature is the crystallization temperature of germanium. With this procedure of calculation the temperature at a given point of the ingot is a function of its value at the end point of the preceding interval.

The procedure was applied in a calculation of the temperature field of germanium single crystals grown under partial shielding on a type MK-1 apparatus (Fig. 1), since the experimental obtained by Verevochkin and Smirnov on the temperature fields of ingots grown on this apparatus enable us to test the validity of the assumptions and the analytical procedure. As apparent from the figure, during growth a portion of the crystal is located in the hot zone, since the temperature of the crucible walls and the nearby shields differ only slightly from the mean temperature of the ingot in the shielding zone. In this zone the heat exchange between the crystal surface and surrounding walls of the crucible may be neglected. We will assume that the heat loss from

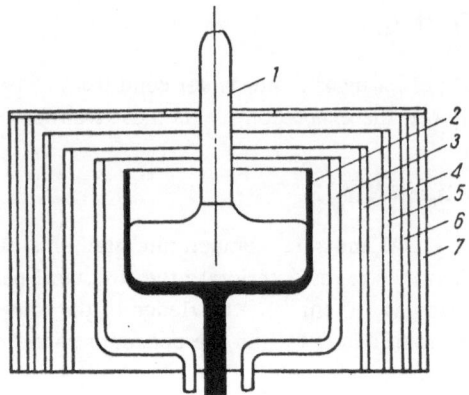

Fig. 1. Arrangement of the shields for the
MK-1 apparatus. 1) Grown crystal; 2)
crucible; 3) heater; 4, 5, 6, 7) shields.

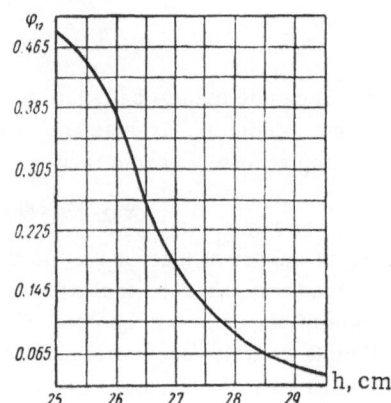

Fig. 2. Variation of the angular
coefficient as a function of position of a given cross section of the
crystal relative to the shields.

the surface of the ingot is attributable to heat exchange with the walls of the apparatus through the spacing between the upper shield and surface of the single crystal.

The main difficulty lies in finding the geometric dependence of the angular coefficients φ_{12} and φ_{21} on
the position of a given cross section of the crystal relative to the shields and crucible walls. The angular coefficient φ_{12} is found from the relation

$$\varphi_{12} = \frac{1}{F_1} \int\limits_{F_1} \int\limits_{F_2} \frac{\cos \varphi_1 \cos \varphi_2}{\pi S^2} \, dF_1 dF_2.$$

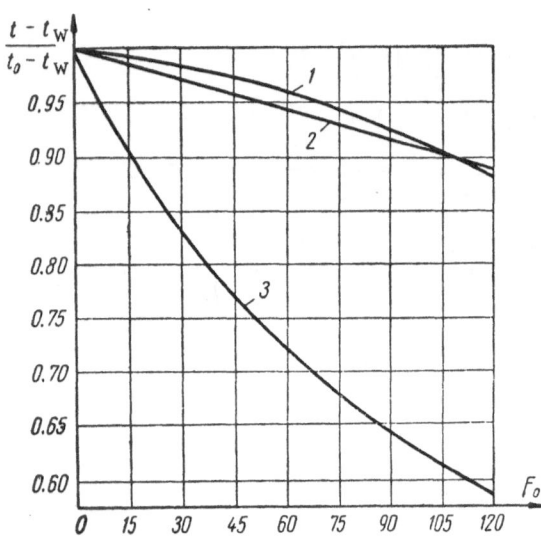

Fig. 3. Relative temperature on the crystal axis in
a cross section near the seed crystal as a function
of the Fourier number ($r_0 = 1$ cm; $v = 1.6$ mm/min).
1) Analytical curve (crystal shielded); 2) experimental curve; 3) analytical curve (no shielding).

The quantity φ_{21} is determined from an analogous expression. In view of the complexity of the integrand,
it is impossible to carry out the integration to completion. Consequently, in order to find the value of the
integral under shielding conditions according to the
arrangement as indicated, numerical methods are in-
voked. A graph showing the dependence of the angular
coefficient φ_{12} on the position of a given cross section
of the crystal relative to the shields for $r_0 = 1$ cm is
presented in Fig. 2.

If the crystal is grown without additional compen-
sation to the melt, the level of the latter will drop continuously in the crucible as the crystal grows. For a
cylindrical crucible the drop in level of the melt $\triangle h$
can be estimated by means of the simple relation

$$\Delta h = \left(\frac{r_0}{R}\right)^2 v\tau, \qquad (11)$$

where R is the radius of the crucible.

For crystals whose length does not exceed 4 cm at
a radius of $r_0 = 1$ cm the reduction in level of the melt

can be neglected. If we are concerned with the temperature field of longer crystals, the drop in level of the melt must be accounted for in determining the heat transfer coefficient.

As noted above, the numerical calculation is affected appreciably by the thermophysical coefficients. The temperature dependence of the thermal conductivity for germanium [8], extrapolated to the melting point from a temperature of 1020°K, was used for the calculations. The thermal conductivity of germanium is virtually independent of the temperature, varying from $0.9 \cdot 10^{-5}$ m^2/sec at 1210°K to $1 \cdot 10^{-5}$ m^2/sec at 700°K, representing a wide range of temperatures.

The results of the calculation were compared with the experimental data of Verevochkin and Smirnov, which were obtained in experiments with the implantation of Chromel–Alumel thermocouples in quartz sheaths within the growing crystal. The analytical curve (Fig. 3) for the temperature field on the axis of a crystal with a radius r_0 = 1 cm suffered a maximum deviation not exceeding 2.0% in the temperature interval from 1050 to 1210°K. Also shown in the figure are data from a calculation of the temperature field on the axis of the ingot under nonshielding conditions. As evident from the graph, shielding of the ingot exerts an appreciable influence on its temperature field.

Literature Cited

1.   E. Billig, Proc. Roy. Soc., Series A, 235:37–55 (1956).
2.   V. L. Indenbom, Macroscopic Theory of the Formation of Dislocations in Crystal Growth, Kristallografiya, No. 2, 594–603 (1957).
3.   P. K. Konakov, V. A. Smirnov, and G. E. Verevochkin, Fundamental Criteria of the Thermal Process in Obtaining Ingots by the Czochralski Method, Coll.: Similarity Theory and Its Application in Heat Engineering (Transzheldorizdat, 1961).
4.   A. V. Lykov, Theory of Heat Conduction (Gostekhizdat, 1952).
5.   N. G. Shimko and P. P. Yushkov, A Finite Hankel Integral Transform, Inzh.-Fiz. Zh., No. 6 (1959).
6.   K. Akijama and I. Yamaguchi, J. Appl. Phys., 33:1899 (1962).
7.   G. A. Slack and C. Glassbrenner, Phys. Rev., 120:782 (1960).

INVESTIGATION OF THE TEMPERATURE FIELD OF
GERMANIUM SINGLE CRYSTALS

G. E. Verevochkin and V. A. Smirnov

In recent times the manufacturers of semiconductor devices have imposed high standards on the quality of the raw material.

In spite of the fact that the properties of germanium as a semiconductor material have been studied the most, the preparation of single crystals with perfect structure and uniform electrical properties is afflicted with all-too-familiar difficulties. One of these is our inadequate knowledge of the laws governing the variation in temperature fields of the crystal and melt.

An attempt is made in the present study to investigate experimentally the temperature field of single crystals prepared by Czochralski method.

On the basis of similarity theory [1, 2] the authors of [3] derived the criteria governing the process. Procedures were developed on this basis for setting up experiments and processing the experimental data, and a semi-empirical formula was derived for calculating the temperature field in germanium single crystals in satisfactory agreement with the experimental results. The tests were carried out in vacuum on the MK-1 laboratory apparatus, as well as in type P17 and Redmet-1 industrial furnaces under actual production conditions. The present article describes an original method for measuring the temperatures in a growing crystal. The resultant formula is compared with the solutions given in [4] and by Uglov in the preceding article.

The MK-1 Experimental Apparatus

For the experimental investigation of the temperature field in germanium single crystals the industrial research technicians of the Moscow Institute of Railroad Engineers (MIIT) fabricated a vacuum furnace, which did not differ in principle from industrial equipment of the same type.

The housing of the vacuum chamber 1 (Fig. 1), together with the top cover 2 and bottom 3 form a hermetically sealed enclosure, inside which the germanium single crystals are grown. A viewing window (4) is provided for visual observation of the processes occurring in the housing 1.

The source of heat for melting the metal is a slot type graphite furnace 5. Two shafts 6 and 7 run through the covers 2 and 3 along the axis of the furnace. The first shaft secures the seed holder 8, the second holds the graphite crucible 9. Both shafts are capable of rotation, as well as displacement along the axis. The construction of the furnace is such that additional shielding can be provided for the growing crystal, for which two current leads 12 and 13 are inserted in addition to the leads 10 and 11 of the primary heater.

In order to minimize the dissipation of heat by the heater 5 it is protected by four graphite shields 15.

All mechanisms of the furnace used to impart motion to the shafts 6, 7 are mounted on a massive iron plate 16.

The upper cover of the furnace is provided with devices to protect the viewing window against overheating and to feed additional alloy to the melt. During the experiments these devices were not used, and the openings through which they normally enter were used for the leads from the ends of the thermocouples.

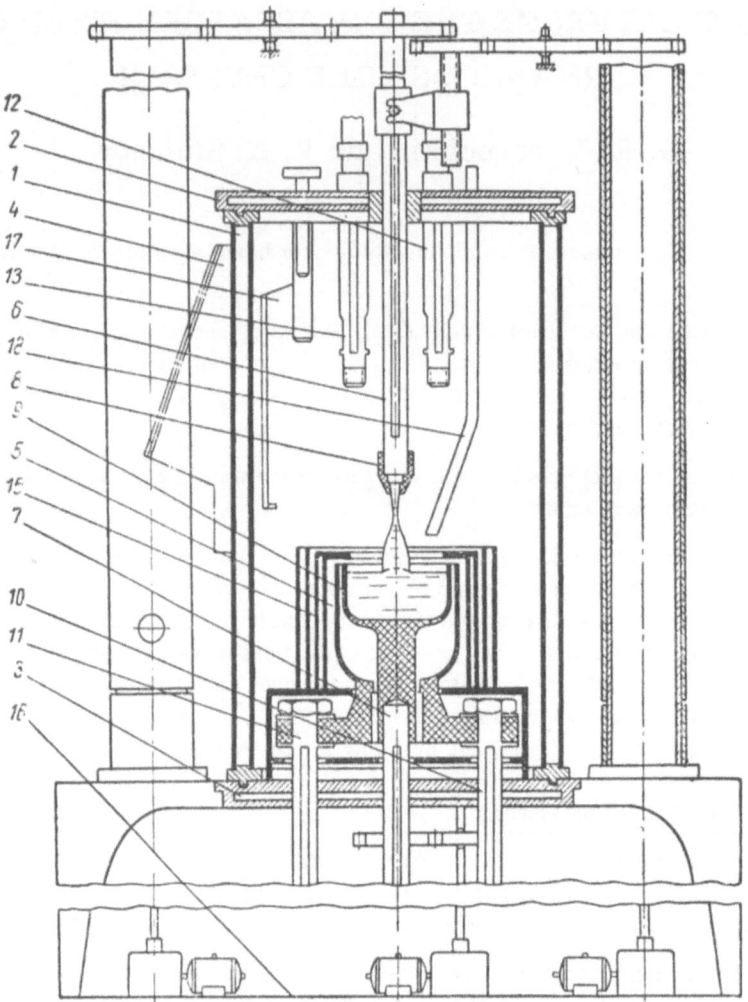

Fig. 1. Schematic diagram of the vacuum furnace.

The housing of the vacuum furnace, the upper and lower cover plates, electrical leads, seed holder, and crucible shafts were cooled throughout the experiments by circulating water. The flow of water through the parts was recorded by type RS-5 flowmeters. A constant pressure of 2 atm was maintained in the cooling system by a sylphon type pressure regulator.

The temperatures of the water coolant at the inlet and outlet of each part was measured with Chromel–Alumel thermocouples throughout the entire pulling process and were recorded with an ÉPP-09M electronic potentiometer.

Power was supplied to the graphite heater from the public outlet through a welding transformer, type ST-34. The voltage on the heater terminals was held constant by a contactless voltage stabilizer. An RU-5-02 master programming device was used for programmed variation of the power input. The velocities of rotation and axial displacement of the seed holder and crucible shafts was measured by means of type TGP-1 tacho-generators.

Preliminary evacuation of the chamber was accomplished by a VN-1 mechanical pump; the final high vacuum was obtained with an N-5S steam-oil pump.

The Experiment and Temperature Measurement Procedure

The graphite crucible 9 (see Fig. 1) was charged with polycrystalline germanium which had been previously etched in aqueous solutions of hydrogen peroxide and hydrochloric acid. Air was then eliminated from the chamber and the metal melted. The seed crystal was lowered well into the body of the melt. At this time the temperature of the melt was made somewhat higher than the crystallization point so that the seed crystal would float fairly freely for better contact with the metal. When the process came to a halt (the seed crystal no longer floating and no more hardening of the melt) the seed crystal was moved upward.

As the crystal grew, the heat balance between the supplied power and dissipation of heat was constantly upset. Stable growth of the ingot was maintained by varying the voltage externally on the heater leads. All voltage variations were recorded throughout the pulling process. Then a power-feed program was formulated and the RU-5 master programming device was used to control the pulling of experimental single crystal specimens with predetermined dimensions.

The first attempt to measure experimentally the temperature field in a single crystal was carried out as follows. It was assumed that if a crystal was pulled from the melt with a known temperature variation at some point, after which the resultant ingot had thermocouples embedded in it and was placed in the furnace, creating as nearly as possible the thermal conditions that existed during pulling, then by the time the temperature at the fixed point reached the value that occurred on segregation of the crystal from the melt, the remaining thermocouples would record the temperature field existing in the specimen at the time of pulling.

For this purpose we pulled germanium single crystals with a diameter of 20–22 mm and a length of 100–120 mm. The fixed point was chosen in the seed crystal, where the temperature was measured during pulling by means of a Chromel–Alumel thermocouple made with a wire 0.2 mm in diameter. The results of the measurements were recorded on the ÉPP-09M electronic potentiometer. Six thermocouples of the same material were embedded in the surface and axis of the resultant specimen.

The single crystals with thermocouples were placed in the furnace, the pressure in the chamber, heater power, and water flow through the various subassemblies were established equal to those at which the crystal segregated from the melt. The crucible was charged with metal, which remained as the ingots were formed. After prolonged heating, the specimen was lowered into the melt. At this instant the temperature in the seed crystal increased slightly, but later in the course of the experiment it remained below the value that occurred in the crystallization process, at 40–50°C. It was concluded on the basis of this that the procedure would yield too coarse an approximation.

The substantial difference in temperature values could be explained by the absence of the steady heat source acting in the pulling process, namely the liberation of latent heat at the crystallization front. Another cause might have been an oxide film formed on the end face of the specimen, thereby creating additional thermal resistance.

Applying the given method, the authors did not propose to obtain a perfect reproduction of the temperature field in the crystal, since the crystallization process is not the same as the melting, but with eradication of the indicated shortcomings it would clearly be possible to obtain satisfactory results. No efforts were undertaken in this direction.

In the laboratory for thermal operation of vacuum systems at MIIT, a procedure was developed and applied for measuring temperatures by growing thermocouples directly in the growing crystal. In our opinion, this approach is extremely ingenious.

Figure 2 shows the construction of a quartz holder used to measure the temperature in the crystal. The ring 1, supported on pins 2, had soldered to it quartz capillaries 3, which served as protective sheaths for the thermocouples. The inside diameter of the ring was 0.5 to 1.0 mm larger than the diameter of the seed crystal shaft and 2 to 3 mm smaller than the diameter of the seed holder.

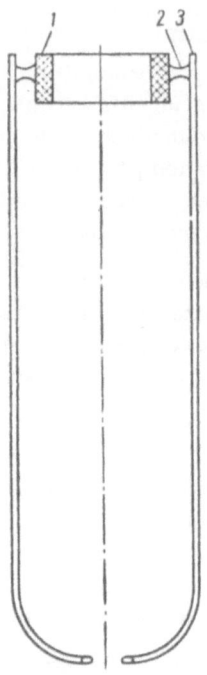

Fig. 2. Construction
of the quartz sheath.

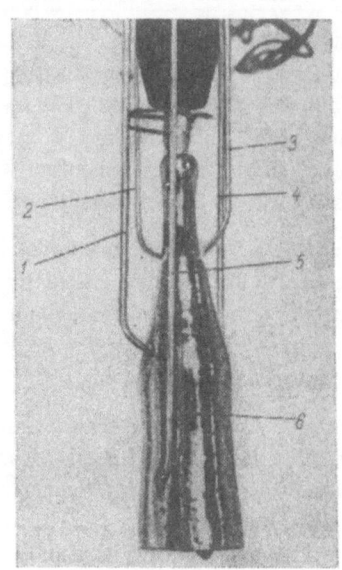

Fig. 3. Germanium single crystal with implanted thermocouples. 1-5) Thermocouples;
6) single crystal.

In order to prevent contact of the thermocouples in the sheath, quartz capillaries 0.3 to 0.35 mm in diameter were threaded at one of the ends. The material for the thermocouples was Chromel–Alumel wire with
d = 0.2 mm. Thus assembled, the thermocouple and quartz sheaths were no more than 1.0 mm in diameter
altogether. The ends of the thermocouples were isolated from contact with the furnace housing by means of a
fiberglass stocking with d = 1.0 mm. Wax residues were removed from the stocking by annealing the latter in
an alcohol burner.

The ends of the thermocouples were led from the interior of the furnace chamber through a rubber vacuum seal in the outlets of the top cover plate, with δ = 8–10 mm. The assembled thermocouples were mounted
on the shaft 6 and suspended on the seed holder 8 (see Fig. 1).

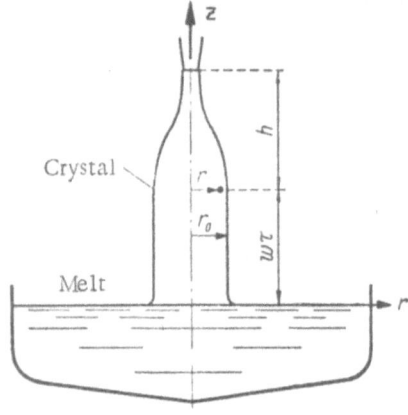

Fig. 4. Diagram showing the process of
single crystal growth.

With this arrangement the process of pulling the single crystal
did not differ in any way from that described previously. The
thoroughly heated seed crystal was lowered into the melt. The
thermocouples were displaced with it until the quartz sheaths rested
on the bottom of the crucible, whereas the shaft for the seed crystal was free to be dropped further down, along with the puller. At
the instant of pulling, when the graphite puller no longer held the
quartz ring, the thermocouples were left on the bottom of the crucible. After the seed holder began to raise the thermocouples, the
latter passed through the molten metal zone, then grew in the
formed single crystal and remained there during the entire experiment. One of the crystals with implanted thermocouples 1–5 is
shown in Fig. 3.

The arrangement of the thermocouples on the bottom of the
crucible prevented us from being able to rotate it during the ex-

periment. A special attachment was built so that the thermocouples could be kept at a distance of 2 or 3 mm from the bottom of the crucible, thus obviating the aforementioned shortcoming.

The thermoemf from the thermocouples was recorded on an ÉPP-09M potentiometer.

A total of about 60 experiments were performed. The germanium ingots were prepared with a diameter of 17–45 mm. The pulling rate was varied within the range 1.4 to 2.5 mm/min. The crucible was rotated at 6 + 10 rpm.

All of the single crystals were rotated in the plane (111). Quartz sheaths 55 to 250 mm in length were used in the experiments.

Processing of the Experimental Data and Discussion

A mathematical investigation of the growth of a germanium single crystal by the Czochralski method (Fig. 4) on the basis of the methods of similarity theory [3] resulted in the following expression for the dimensionless temperature field:

$$\frac{t}{t_0} = \varphi\left(F_0, \ Pe, \ \frac{\sigma_0 T_0^3 r_0}{\lambda_0}, \ \frac{h}{r_0}, \ \frac{r}{r_0}\right), \tag{1}$$

where τ is the time (reckoned from the initial time τ_0 when a given cross section first emerges from the melt; τ_0 = h/w), t is the instantaneous value of the temperature in the single crystal, t_0 is the crystallization temperature of·germanium, h is the coordinate of the observed point on the z axis, r is the coordinate of that point on the radius r, r_0 is the mean radius of the ingot, w is the crystal growth rate, a_0 and λ_0 are the coefficients of thermal diffusivity and thermal conductivity of germanium at the crystallization temperature, σ_0 is the black body radiation constant, Fo = $a_0\tau/r_0^2$ is the Fourier number, Pe = wr_0/a_0 is the Péclet number, $\sigma_0 T_0^3 r_0/\lambda_0$ is the radiant heat transfer criterion.

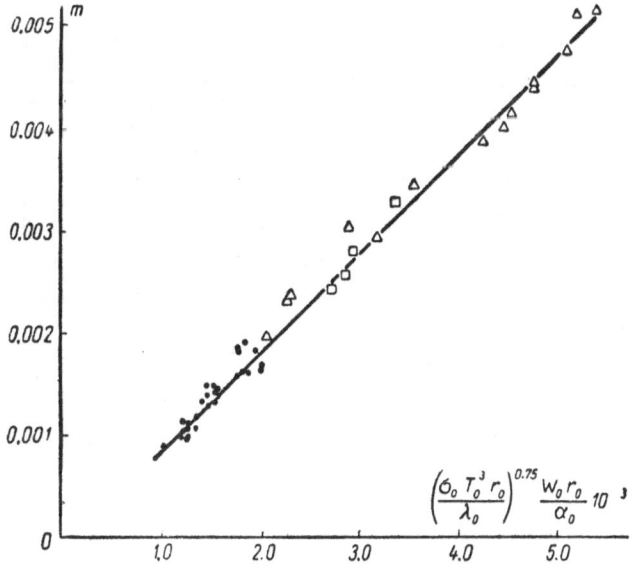

Fig. 5. Processing of experimental data obtained on the following apparatus: ● — MK-1; △—P-17; □ —Redmet-1.

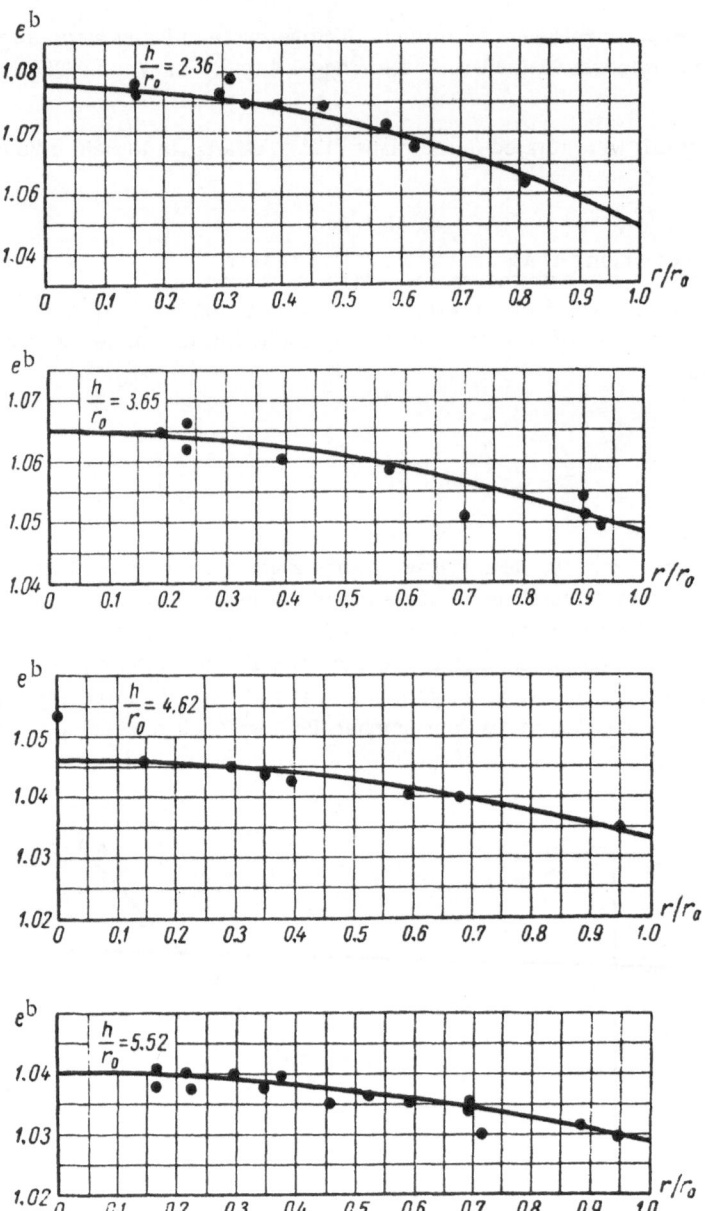

Fig. 6. Graph of the

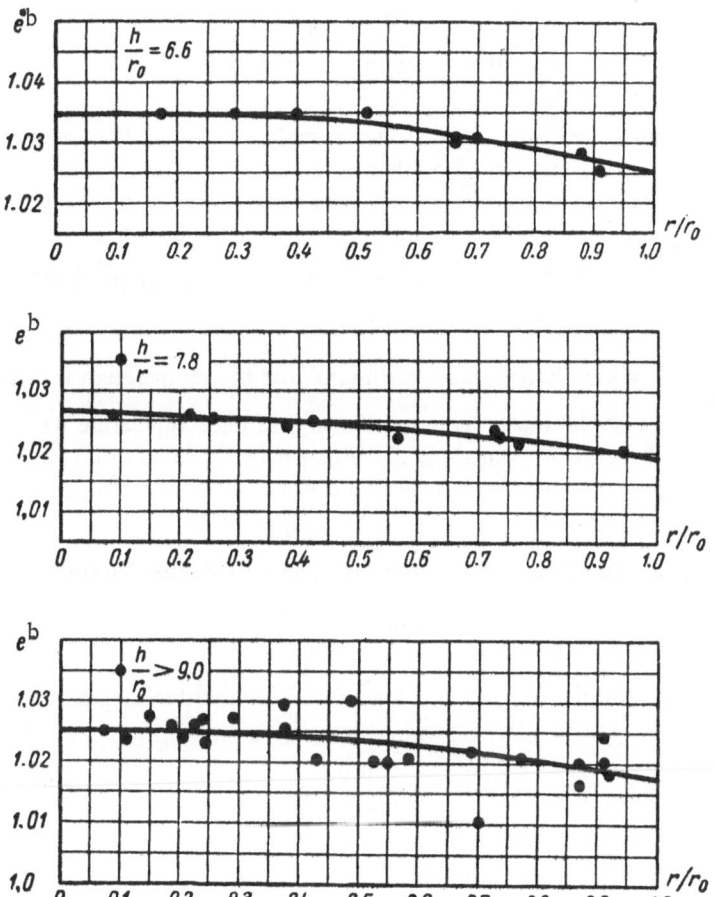

dependence $e^b = \varphi(r/r_0)$.

Processing of the experimental data according to Eq. (1) showed that the function $\ln t/t_0$ qualitatively does not depend on the ratio h/r_0 or r/r_0 and depends linearly on the Fo number for a crystal of constant diameter, i.e., the cooling of the ingot is typified by regular behavior. For each melting we constructed graphs of the dependence $\ln t/t_0 = \psi(\text{Fo})$. These graphs were represented analytically by the relation

$$\ln t/t_0 = b - m\,\text{Fo}. \qquad (2)$$

It is clear from an analysis of the experimental data, as well as the physical interpretation of the radiation effect, that the quantity

$$m = \frac{\ln\left(\frac{t_1}{t_0}\right) - \ln\left(\frac{t_2}{t_0}\right)}{\text{Fo}_2 - \text{Fo}_1}, \qquad (3)$$

which characterizes the cooling rate of the ingot, cannot remain constant as the conditions of heat exchange between the crystal and the surrounding furnace walls vary. Such a variation occurred when ingots of different diameters were pulled, or when they were pulled at faster or slower rates. Other factors could not affect the value of m, since the furnace geometry, heater, and shields were not changed in the course of the experiments.

A graph of the cooling rate m as a function of the dimensionless group $(\sigma_0 T_0^3 r_0/\lambda_0)\,\text{Pe}$ is shown in Fig. 5.

The quantity e^b is shown in Fig. 6 as a function of the variable r/r_0. It is evident from inspection of $e^b = \dfrac{t\,t_0}{e^{-m\text{Fo}}}$ as a certain temperature simplex that the temperature is distributed in the crystal cross section according to a parabolic law. The nonuniformity of the temperature field decreases as the relative length h/r_0 of the crystal increases. However, this effect is observed up to a value of $h/r_0 \cong 8$. A further increase in the relative length of the ingot to $h/r_0 = 14.0$ does not affect the temperature distribution in the crystal.

The graph in Fig. 6 was represented in the following general analytical form:

$$e^b = c - a\left(\frac{r}{r_0}\right)^2. \qquad (4)$$

The coefficients a and c of the parabola as a function of the relative length of the crystal were approximated by the equations

$$a = \frac{0.06}{\dfrac{h}{r_0}}, \qquad (5)$$

$$c = 1.135 - 0.0247\,\frac{h}{r_0} + 0.00137\left(\frac{h}{r_0}\right)^2 \qquad (6)$$

in the interval $h/r_0 \le 8.0$; for $14.0 > h/r_0 \ge 8.0$ we have $a = 0.006$ and $c = 1.025$.

Substituting the values of b into Eq. (2), we obtain an expression for calculating the temperature field of the single crystal:

$$\frac{t}{t_0} = \left[c - a\left(\frac{r}{r_0}\right)^2\right]e^{-m\text{Fo}}. \qquad (7)$$

A comparison of the experimental values of the dimensionless temperature with the values calculated according to Eq. (7) is made in Fig. 7. The limiting relative error does not exceed 4%.

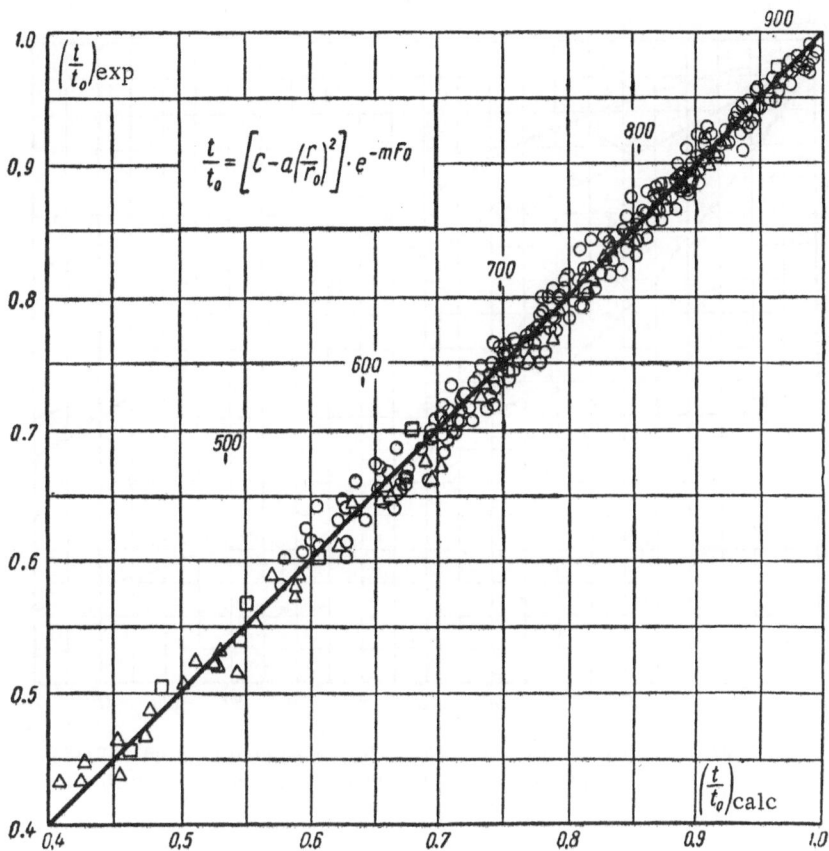

Fig. 7. Experimental and analytical values of the dimensionless temperature as obtained: ● —on the MK-1 apparatus; △—on the P-17 apparatus; ◻ —on the Redmet-1.

Differentiation of Eq. (7) with respect to the variables r and h yields the following equations for the radial and axial temperature gradients:

$$\frac{dt}{dr} = -\,0.12 t_0\,\frac{r}{hr_0}\,e^{-mFo};$$

(8)

Table of Temperature Gradients

τ	$h=3.0$ cm		$h=5.0$ cm		$h=7.0$ cm		$h=9.0$ cm	
min	$\mathrm{grad}_r\,t$	$\mathrm{grad}_h\,t$	$\mathrm{grad}_r\,t$	$\mathrm{grad}_h\,t$	$\mathrm{grad}_r\,t$	$\mathrm{grad}_h\,t$	$\mathrm{grad}_r\,t$	$\mathrm{grad}_h\,t$
12.5	32.9	52.4	19.8	48.5	14.15	50.6	11.0	53.3
25	29.7	46.6	17.8	43.8	12.8	45.7	9.9	48.0
37.5	26.4	41.5	15.8	38.9	11.3	40.6	8.8	42.7
50	23.5	37.0	14.15	34.7	10.25	36.2	7.8	38.0
62.5	20.9	32.8	12.5	30.8	8.9	32.1	6.95	33.8
75	18.6	29.2	11.2	27.4	8.0	28.6	6.2	30.2

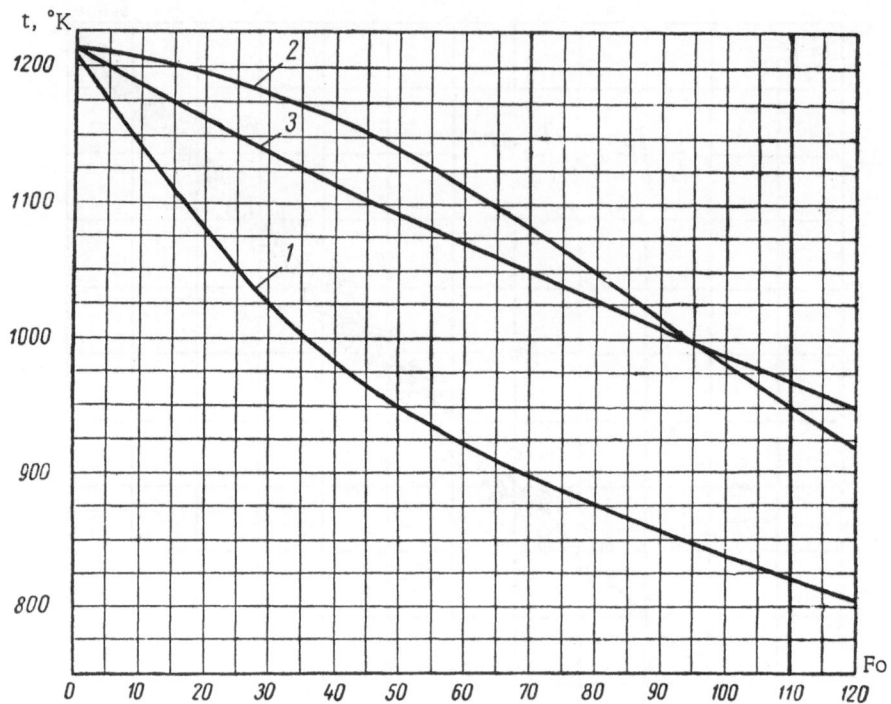

Fig. 8. Numerical calculations of the temperature distribution in a germanium crystal. 1) For pulling of an unshielded crystal; 2) for partial shielding; 3) according to Eq. (7).

$$\frac{dt}{dh} = t_0 e^{-m\mathrm{Fo}} \frac{ma_0}{wr_0^2} \left[1.135 - 0.0247 \frac{h}{r_0} + 0.00137 \left(\frac{h}{r_0} \right)^2 - \right.$$
$$\left. - 0.06 \frac{r^2}{hr_0} - \frac{wr_0^2}{ma_0} \left(0.0247 - 0.00274 \frac{h}{r_0} - 0.12 \frac{r^2}{h^2} \right) \right], \tag{9}$$

for the interval $h/r_0 \leq 8.0$ and

$$\frac{dt}{dh} = t_0 e^{-m\mathrm{Fo}} \frac{ma_0}{wr_0^2} \left[1.025 - 0.007 \left(\frac{r}{r_0} \right)^2 \right] \tag{10}$$
$$\text{for} \quad 14 \geqslant \frac{h}{r_0} \geqslant 8.0.$$

For a germanium ingot with r = 1.0 cm grown at the rate w = 1.6 mm/min the radial and axial gradients in cross sections at h = 3, 5, 7, and 9 cm are shown in the table.

Making use of the value of the thermal conductivity of germanium [5] reduced to the mean temperature of the ingot surface, we calculated the heated flux from the surface of a crystal with d = 30 mm and l = 200 mm, for which the heat of crystallization was about 25% of the total flux.

It is instructive to compare the values of the temperatures obtained according to Eq. (7) and the experimental data with the theoretical solutions of Billig [4] and Uglov (preceding article).

In comparing Billig's solution with Eq. (7), the quantities λ and ε were normalized to t = 800°C.

Values of λ = 24.5 W/m°C [5] and ε = 0.57 (see p. 147) were assumed for germanium.

Inasmuch as Billig's solution applies to a cylindrical ingot of infinite length in the positive z direction, the calculations based on Eq. (7) were carried out for the case when $h/r_0 > 9.0$, i.e., for the interval in which the relative length of the crystal does not affect the temperature distribution.

The disparities between the theoretical solutions and the data obtained from Eq. (7) and experiment attain 40%. This may be attributed to the fact that Billig's solution was found for the case of free radiation from the crystal; the experimental data and data from Eq. (7) were derived under actual pulling conditions, where free radiation is always impeded by the heated regions of the furnace. A certain error clearly lies in the fact that Billig's solution does not take into account the latent heat of crystallization of the material.

The results of a comparison of a numerical calculation of the temperatures in a germanium crystal (according to Uglov) with a calculation based on Eq. (7) are shown in Fig. 8.

If we consider that the discrepancy in the analytical dependence against the experimental data lie within 4% limits, it may be stated with assurance from inspection of the graphs that the calculation of the temperature field, even taking into account partial shielding, is not distinguished by high accuracy.

In the present work we have: a) proposed a method for measuring the temperatures in single crystals grown from the melt by the Czochralski method; b) derived a semiempirical equation for the temperature distribution in single-crystal ingots in satisfactory agreement with the experimental data; c) performed an analysis of the variation in radial and axial temperature gradients in germanium single crystals.

Literature Cited

1.   M. V. Kirpichev, Similarity Theory (Izd. Akad. Nauk SSSR, 1953).
2.   P. K. Konakov, Similarity Theory and Its Application in Heat Engineering (Gosénergoizdat, 1960).
3.   P. K. Konakov, V. A. Smirnov, and G. E. Verevochkin, Fundamental Criteria of the Process of Obtaining Ingots by the Czochralski Method, Tr. MIITa (Transactions of the Moscow Railroad Engineers Institute), No. 139 (1961).
4.   E. Billig, Proc. Roy. Soc., Series A, 235:37–55 (1956).
5.   A. Slack and C. Glassbrenner, Phys. Rev., 120:(3):(1960).

EXPERIMENTAL INVESTIGATION OF THE SHIELDING CONDITIONS AND THE NEUTRAL MEDIUM ON THE TEMPERATURE FIELD OF A CRYSTAL GROWN FROM THE MELT

L. A. Zaruvinskaya and V. A. Smirnov

The trend toward production of single crystals with a high degree of perfection on the part of the crystal lattice and uniform impurity distribution throughout the entire volume of the crystal has resulted in a variety of improvements in the technological process of growing crystals by the Czochralski seed-pulling technique. In the industrial environment these improvements have in recent times led to the design of numerous systems for shielding the melt and crystal, as well as the application of various media in which to grow the crystals. This is no mere chance, because the preparation of a single crystal perfect in its structure and attributes depends largely on the thermal process involved in its growth.

Despite the many references in the literature to the connection between the structure and properties of crystals grown from the melt and thermal factors, only a very few papers [1–3] are devoted to the theoretical and experimental investigation of the thermal processes in crystal growth. This is unquestionably related to the difficulties encountered both in the analytical solution and in the setting up of experiments to ascertain the temperature fields in the nonstationary process that transpires at high temperatures.

In the preceding article Verevochkin and Smirnov have proposed an original method for investigating the temperature field of a crystal during its growth, and they have cited a semiempirical equation for calculating the temperatures in a germanium crystal as it grows in vacuum without the use of special shielding. Francois [3] has published the only paper in which the problem of selecting a shielding system for the preparation of crystals devoid of thermal stresses is solved theoretically. In this paper an analysis is made of the influence of various types of shielding and neutral media on the temperature field of a germanium crystal prepared by the Czochralski method.

The difficulty of an analytical calculation of the nonstationary temperature field in a crystal under the conditions of arbitrary shielding and the impossibility of a proper theoretical evaluation of the heat transfer coefficient in the growth of a crystal in an inert medium have elicited the need for an experimental investigation of the effect of various modes of shielding and neutral media on the temperature field and quality of the single crystal.

More than sixty experiments have been run in industrial furnaces and in laboratory equipment at high vacuum (0.0133 N/m^2) and in an argon atmosphere (with a pressure differential of $0.294 \cdot 10^5$ N/m^2), as well as in an argon-hydrogen mixture with well-defined seed-pulling parameters and various conditions of shielding the crystal and melt. The results of temperature measurements in the ingots are compared with the temperature field of a crystal grown in vacuum without special shielding.

To determine the temperature fields of the crystals during their growth we used a procedure developed at the laboratory for thermal operation of vacuum systems of the Moscow Institute of Railroad Engineers (MIIT). Chromel–Alumel thermocouples encased in quartz sheaths were introduced into the melt and held there by a special attachment. After being grown right inside the single crystal as the latter formed, the thermocouples and crystal were raised during the growth process, passing in sequence the different shielding zones. The emf of the thermocouples was continuously recorded by an ÉPP-09 electronic potentiometer, thus making it possible

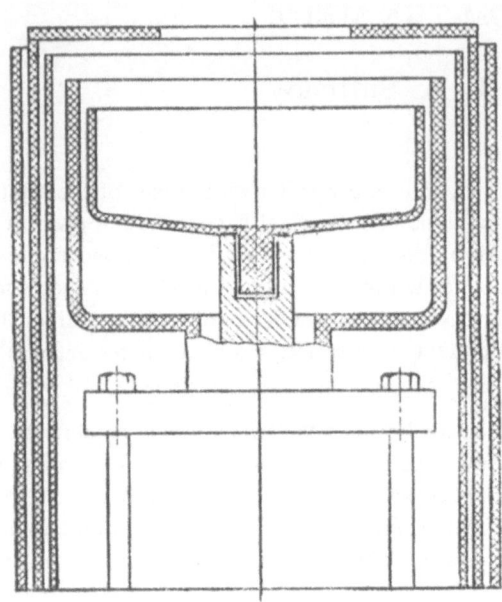

Fig. 1. Diagram of the industrial apparatus with standard shielding arrangement.

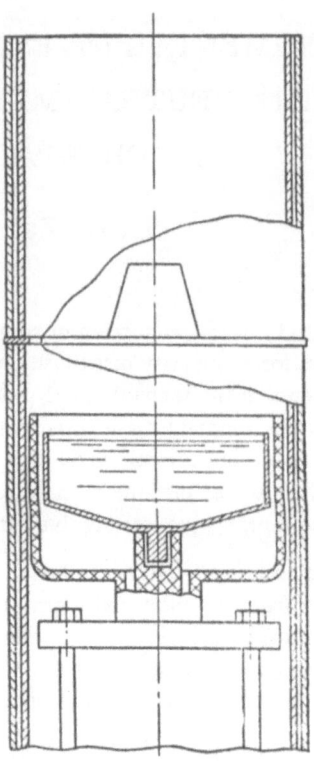

Fig. 2. Diagram of industrial apparatus with upper graphite shield.

to trace the variation in temperature within the crystal at various distances from the level of the melt. The temperature of the graphite shields was determined simultaneously by means of Chromel–Alumel thermocouples or an optical pyrometer.

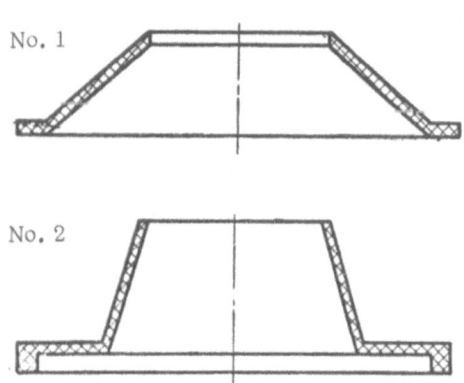

Fig. 3. Graphite shields floating on the surface of the melt.

The arrangements and geometry of the shields used in the experiments are illustrated in Figs. 1 and 2.

Melting in the industrial apparatus (Figs. 1 and 2) was performed both in vacuum (0.0133 N/m^2) and in an inert gas atmosphere (argon at a pressure of $0.294 \cdot 10^5$ N/m^2), with a standard shielding arrangement and with the shields most frequently used under industrial conditions in these furnaces (Fig. 3). The seed-pulling conditions also conformed to industrial practice; the mean rate of ascent of the seed crystal was 1.8 mm/min, the crucible turned at a rate of 8–10 rpm, until the instant of growth of the thermocouples the seed rotated at 20–30 rpm. After implantation of the thermocouples the ingot was no longer rotated.

Preliminary experiments showed that with the cessation of rotation of the crucible the ingot tended to "run" toward the colder part of the melt, and the presence of slag on certain portions of the surface of the melt aggravated distortion

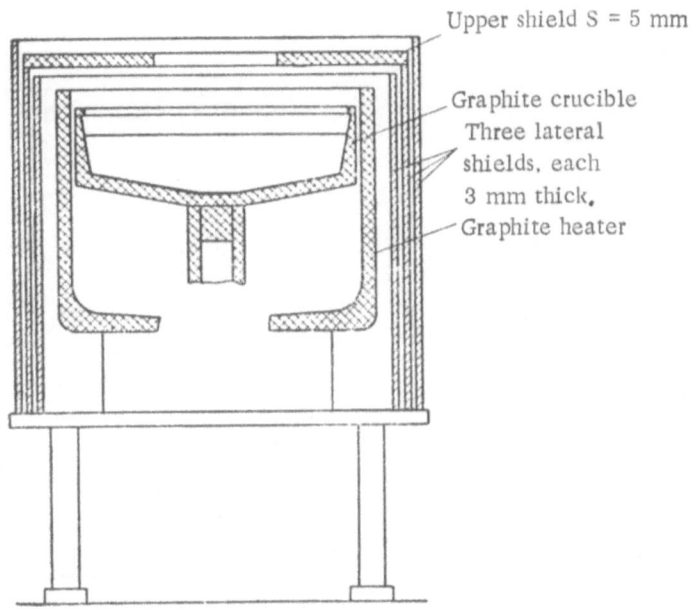

Upper shield S = 5 mm

Graphite crucible
Three lateral
shields, each
3 mm thick.
Graphite heater

Fig. 4. Industrial apparatus with standard shielding arrangement.

of the ingot geometry. Stopping rotation of the ingot, on the other hand, did not impose significant changes in the geometry of the crystal and did not violate its single-crystal character. A slight change in the diameter of the ingot (1 or 2 mm) when its rotation was stopped could be compensated by smooth regulation of the heater power. So as not to elicit any change in diameter, most of the experiments were carried out without rotation of the seed crystal. No deterioration of the single crystals was observed with this practice.

The metal charge in the crucible amounted to 3 or 4 kg. Cylindrical ingots were grown, 30–35 mm in diameter and with slight conical swelling at the seed. The ingots reached a length of 250–300 mm.

The temperature fields of crystals grown with the various shielding arrangements were determined in the industrial apparatus (Fig. 4). Several types of shields floating on the surface of the melt and shielding the surface of the ingot at a height of 40–60 mm from the level of the melt were investigated. The shields had a high sloping conical section, with an opening large enough in diameter for the ingot to emerge. Experiments were also conducted for comparison without the floating shields, just with the overhead shield used in the standard shielding arrangement. The experiments were carried out in vacuum (0.0133 N/m²), as well as in an argon atmosphere ($0.294 \cdot 10^5$ N/m²). The seed-pulling conditions conformed to industrial practice (the mean rate of ascent of the seed was 1.8 mm/min; the crucible was rotated at 8–10 rpm; the seed crystal, as a rule, was not rotated). With a 2.5 kg charge of metal in the crucible ingots were grown with an approximately uniform diameter of 25 to 28 mm and length of 200 to 250 mm.

In addition, the temperature field of a dislocation-free crystal (or crystal with low dislocation density of no more than 100 cm⁻²) was investigated in the apparatus. The crystals were prepared under industrial conditions using a special floating shield with a quartz sleeve (Fig. 5a) in an atmosphere of argon and hydrogen with the mixture at a pressure of $0.343 \cdot 10^5$ N/m². For comparison the temperature of the crystal was determined under the same growing conditions but with a graphite sleeve of similar geometry (Fig. 5b). We were unable with this mode of shielding to obtain a dislocation density lower than 10^2 to 10^3 cm⁻².

Experiments were performed in the laboratory apparatus (Fig. 6) with the growing of germanium single crystals with heated and "cold" shields at high vacuum (0.0133–0.00133 N/m²). The seed-pulling parameters

a

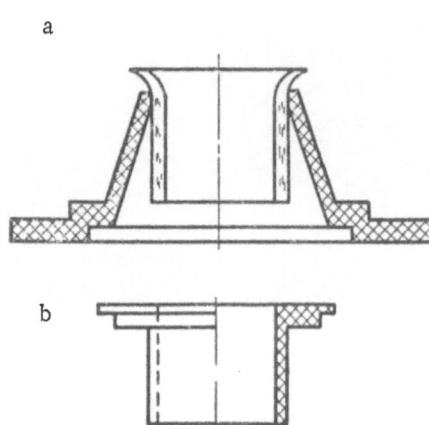

b

Fig. 5. Shielding employed for the preparation of dislocation-free crystals. 1) Floating graphite shielded with quartz sleeve; b) graphite sleeve.

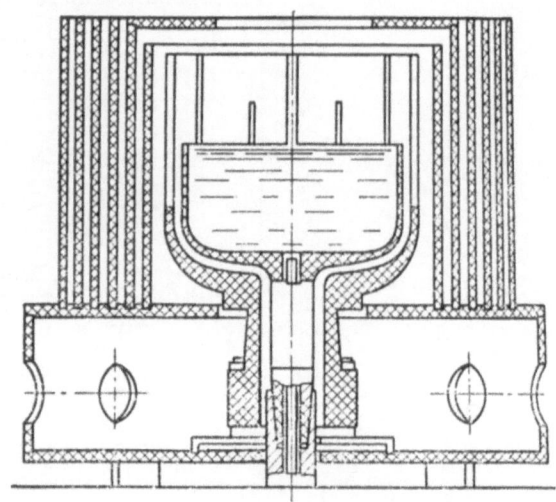

Fig. 6. Laboratory apparatus with standard shielding arrangement.

View ABC

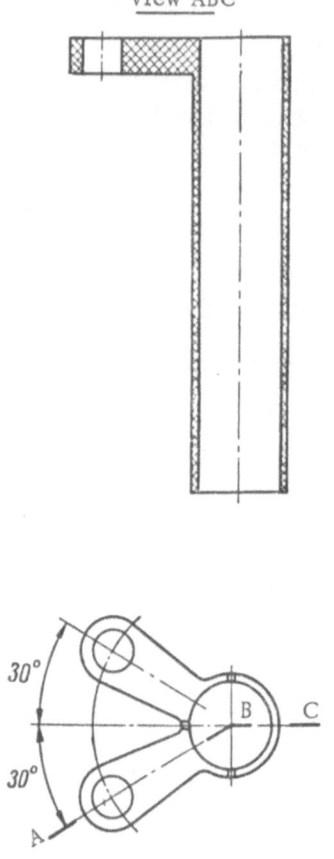

Fig. 7. Heated shield of laboratory apparatus.

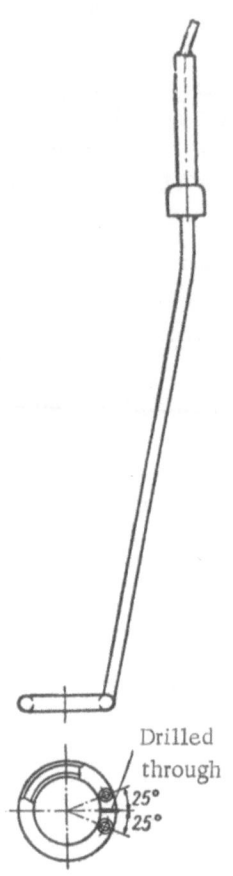

Fig. 8. Cooled shield of laboratory apparatus.

included: mean rate of ascent of seed 1.6 mm/min, rotation of crucible at 6–7 rpm, rotation of seed crystal at 15–20 rpm. With a 550 g charge of metal in the crucible-crystals were grown with a diameter of 20 to 23 mm and a length of 120 to 170 mm.

The heated shield (Fig. 7) was a hollow graphite cylinder 42 mm in diameter with a wall thickness of 3 mm. The lateral surface of the shield had four slots cut through it, 1.5 mm in width. The upper part of the shield was provided with lugs for connection to electrical leads installed in the top cover of the apparatus. The shield was 200 mm in length, which permitted shielding of the ingot almost over its entire length. The shield was centered with respect to the axis of the seed crystal and was secured at a distance of 20 mm from the level of the melt. With a power of about 3 kW supplied to the main heating unit, a current of 50–60 A was delivered to the heated shield at a voltage of 10–12 V, which meant that the temperature of the shield (measured with an OPPIR-017 optical pyrometer) could be maintained at 830–870°C. Any attempt to raise the temperature of the shield or to bring it near the surface of the melt caused floating of the seed crystal and prevented controlled selection of the nucleation and growing conditions. The pulling process did not begin until the entire surface of the shield had acquired a uniform temperature. Thermocouples encased in quartz sheaths were inserted inside the shield and were immersed in the melt as soon as they were caught by the growing crystal.

Several experimental meltings were made with a "cold" shield (Fig. 8), consisting of a stainless steel ring with an inside diameter of 39 mm and wall thickness of 1.0 mm, cooled by water. The ring was centered with respect to the axis of the seed crystal and could be placed at different levels above the surface of the melt.

All of the experiments were conducted on zone-refined germanium doped with antimony. The resistivity of the grown single crystals was from 3 to 10 $\Omega \cdot$ cm. In each series of tests sample investigations were run on the quality of the crystals, the dislocation density being evaluated in two or three cross sections of the ingot, along with the resistivity at the ends and sides parallel to the axis. In some cases anodic etching was performed on longitudinal and transverse sections of the ingot.

The shields (except for the "cold" ring for the laboratory apparatus), crucibles, and heaters for all the furnaces were made of grade GM graphite.

Processing of the Experimental Data

In the preceding article Verevochkin and Smirnov analyzed the regularity of the thermal conditions associated with cooling of a crystal during its growth from the melt, wherein they showed that the temperature field in the crystal is determined by the criterion Fo = $a_0 \tau / r_0^2$, the thermal characteristics of the crystal, and the conditions of heat exchange with the surrounding medium, as well as the pulling rate and geometry of the crystal.

We processed the experimental data in our own work in conformity with this analysis. On completion of the experiment the geometry of the grown crystal was determined: its length l, mean radius r_0, height h of the ingrown thermocouple relative to the seed, radius r of the ingrown thermocouple. The pulling time τ and length l were used to determine the mean growth rate w = l/τ, which differed from the rate of ascent of the seed due to dropping of the level of the melt during growth. The thermogram data were presented as a function of the time criterion. To do this, the temperature values were copied from the thermograms obtained with the ÉPP-0.9 potentiometer for a point with h/r_0 and r/r_0 as a function of the time segregation τ_0 of the given cross section from the melt, and graphs of

$$\ln t/t_0 = \varphi \, (\text{Fo}),$$

were constructed, where t is the instantaneous temperature at the given point, °C; t_0 is the crystallization temperature (assumed equal to 936°C for germanium).

It is apparent from the graphs (Figs. 9–11) that the dependences are straight lines whose slope, according to the theory of thermal regularity, characterizes the cooling rate m of a given cross section of the ingot. The value of m depends on the shape and thermal characteristic of the ingot, as well as the conditions of heat exchange with the surrounding medium.

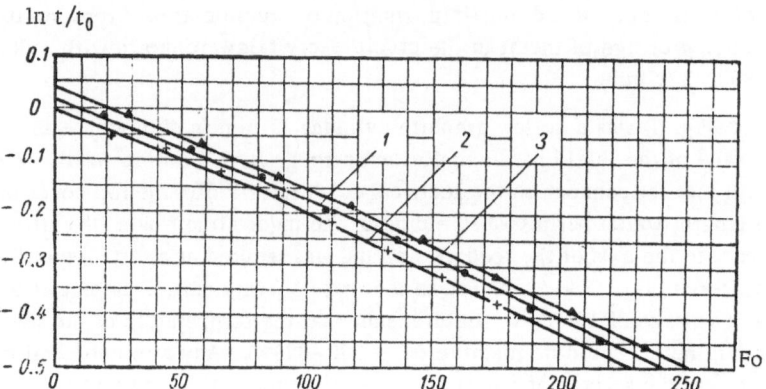

Fig. 9. Graph of the dependence $\ln t/t_0 = \varphi(Fo)$; m = 0.0023; m_{sh} = 0.001. 1) Melting in vacuum with standard shielding; 2) melting in vacuum with floating shield No. 1 (Fig. 3a); 3) melting in vacuum with floating shield No. 2 (Fig. 3b).

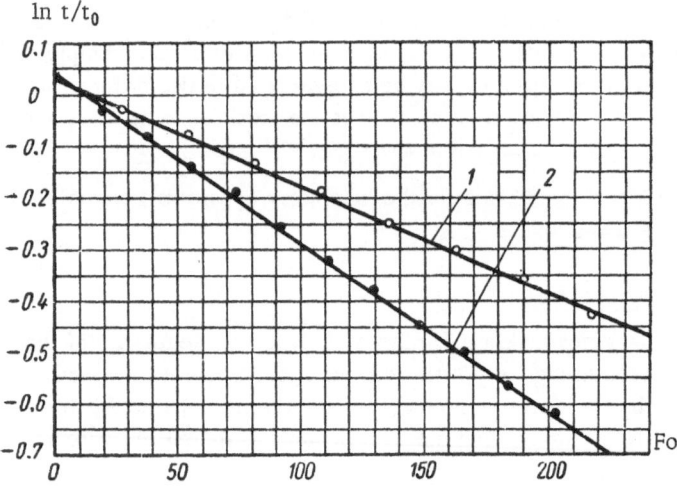

Fig. 10. Cooling rates in the growth of crystals in vacuum and in an argon atmosphere. 1) Melting in argon with standard shielding (m = 0.0021); 2) melting in vacuum with standard shielding (m = 0.000325).

In the preceding article a graph is constructed to show the relation

$$m = f\left(\frac{\sigma_0 T_0^3 r_0}{\lambda_0}, \quad \frac{w r_0}{a_0}\right),$$

where σ_0 = 5.7 W/m²-°K is the black body radiation coefficient, T_0 is the crystallization temperature, λ_0 is the thermal conductivity at the crystallization temperature (assumed equal to 33.5 W/m°C for germanium), a_0 is the thermal diffusivity at the crystallization temperature (assumed equal to 16.5 · 10^{-6} m²/sec for germanium).

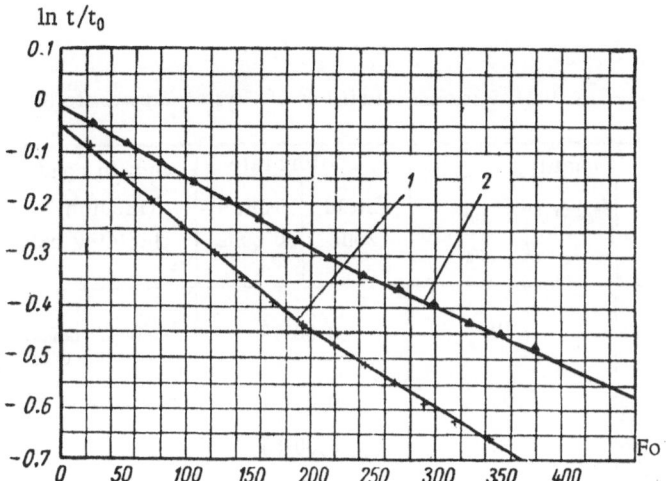

Fig. 11. Graph of the dependence $\ln t/t_0 = \varphi(Fo)$. 1) Melting in argon with floating shield No. 1 ($m = 0.0011$; $m_{sh} = 0.00135$); 2) melting in argon with floating shield No. 2 ($m = 0.0015$; $m_{sh} = 0.0020$).

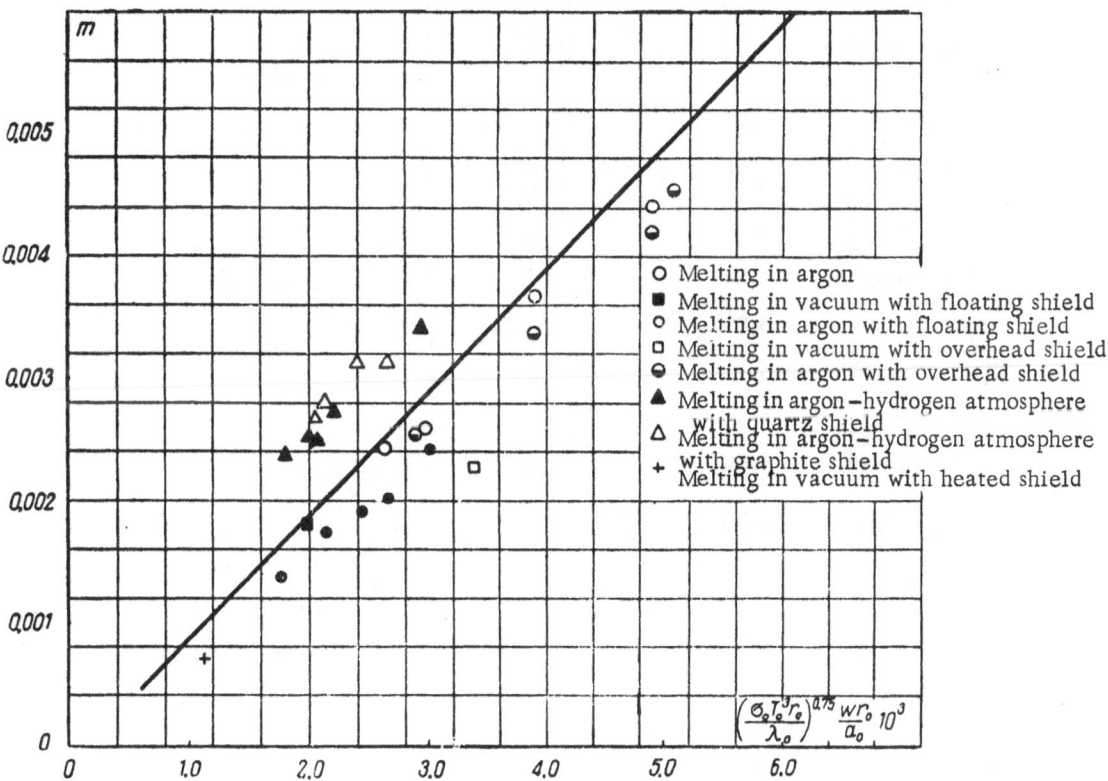

Fig. 12. Cooling rates of crystals grown in various media and modes of shielding (the line represents the dependence $m = \varphi\left(\dfrac{\sigma_0 T_0 r_0}{\lambda_0}, \dfrac{wr_0}{a_0}\right)$ for vacuum with the standard shielding arrangement).

Cooling Rates m Under Various Conditions of Crystal Growth from the Melt

Test No.	Quantity			Cooling rate		Growth conditions (medium and shielding mode)
	h/r_0	r/r_0	$\left(\dfrac{\sigma_0 T_0^3 r_0}{\lambda_0}\right)^{0.75}\dfrac{wr_0}{a_0}\cdot 10^3$	shield zone	unshielded zone	
1	7.62	0.743	5.19	0.00518		Vacuum, standard shielding (Fig. 1)
2	2.93	0.326 / 0.604	5.09	0.00481		The same
3	13.21	0.485 / 0.688	4.52	0.00455		»
4	8.92	0.216 / 0.730	4.45	0.00426		»
5	11.42	0.209 / 0.552	4.26	0.00399		»
6	8.81	0.105	3.18	0.00300		»
7	9.84	0.369	2.88	0.00306		»
8	13.48	0.677	2.36	0.00266		»
9	12.00	0.513	2.76	0.00291		»
10	10.90	0.378	2.39	0.00270		»
11	5.20	0.213	3.24	0.00322		»
12	6.78	0.318	4.90	0.00442		Argon, standard shielding (Fig. 1)
13	14.25	0.316 / 0.421	3.91	0.00366		The same
14	5.28	0.497	2.96	0.00255		»
15	5.56	0.262	2.64	0.00240		»
16	12.60	0.740	2.05	0.00210		»
17	10.20	0.325	3.38	0.00227		Vacuum, overhead shield (Fig. 2)
18	10.20	0.437	2.89	0.00250		Argon with overhead shield (Fig. 2)
19	13.20	0.344 / 0.888	3.93	0.00333		The same
20	7.10	0.476 / 0.667	5.1	0.00454		»
21	11.50	0.333 / 0.905	4.88	0.00417		»
22	8.00	0.659	1.815	0.00170	—	Vacuum, floating shield No. 2 (Fig. 3)
23	14.15	0.635	1.73	0.00170	—	The same
24	13.10	0.592 / 0.890	2.235	0.00170	0.00230	Vacuum, standard shielding, d = 60 mm (Fig. 4)
25	9.26	0.567	2.442	0.00170	0.00230	Vacuum, floating shield No. 1 (Fig. 4)
26	12.30	0.462	1.975	0.00170	0.00230	Vacuum, floating shield No. 2 (Fig. 4)

Cooling Rates m Under Various Conditions of Crystal Growth from the Melt

Test No.	Quantity			Cooling rate		Growth conditions (medium and shielding mode)
	h/r_0	r/r_0	$\left(\dfrac{\sigma_0 T_0^3 r_0}{\lambda_0}\right)^{0.75}\dfrac{wr_0}{a_0}\cdot 10^3$	shield zone	unshield-ed zone	
27	5.30	0.635	3.00	0.00235	0.00190	Argon, floating shield No. 1 (Fig. 4)
28	5.70	0.880	2.61	0.00200	0.00150	Argon, floating shield No. 1 (Fig. 4)
29	4.74	0.970	1.77	0.00135	0.00110	Argon, floating shield No. 2 (Fig. 4)
30	14.80	0.427	2.89	0.00310	—	Argon-hydrogen mixture, shielding for dislocation-free crystal, w = 2 mm/min Fig. 4
31	9.17	0.192 0.654	2.30	0.00210	—	The same
32	13.84	0.320 0.763	2.24	0.00250	—	»
33	15.50	0.742	2.03	0.00200	—	»
34	14.70	0.318 0.682	1.99	0.00250	—	Argon-hydrogen mixture, shielding for dislocation-free crystal, w = 1.6 mm/min (Fig. 4)
35	11.70	0.625	2.05	0.00265	—	Argon-hydrogen mixture, shielding with graphite sleeve, w = 2 mm/min (Fig. 4)
36	7.50	0.057 0.543	2.575	0.00311	—	The same
37	10.10	0.000 0.584	2.140	0.00275	—	»
38	10.60	0.230 0.682	2.420	0.00312	—	»
39	7.70	0.520	1.800	0.00285	—	The same, w = 96 mm/min
40	12.90	0.330	1.13	0.00069	—	Vacuum, heated shield with temperature t = 850°C (Fig. 6)

As evident from the graph (Fig. 12), for crystals grown in vacuum without special shielding, m forms a straight line dependence. Also shown on the graph for comparison are the values of m for various shielding modes and media investigated in the present study.

The values obtained for the cooling rate m for the various conditions of single crystal growth are compiled in the table.

The temperature in the crystal was calculated according to the semiempirical formula proposed by Verevochkin and Smirnov in the preceding article:

$$\frac{t}{t_0} = \left[c - a\left(\frac{r}{r_0}\right)^2\right] e^{-m\,\text{Fo}},$$

where a and c are empirical coefficients, functions of the parameter h/r_0.

This formula was derived for the case of crystal pulling in vacuum with the standard shielding arrangement, for a fairly large-diameter opening in the overhead shield (77 and 90 mm).

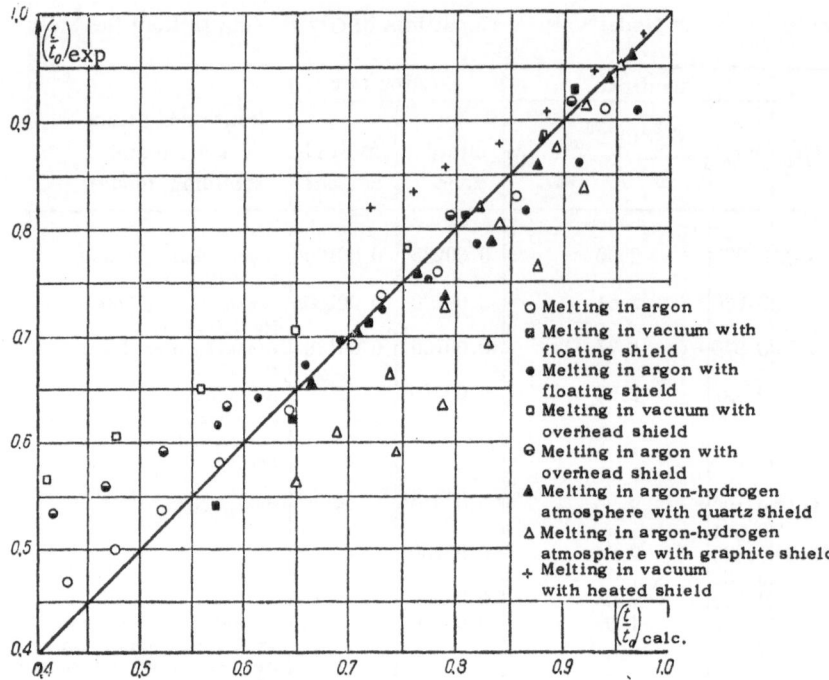

Fig. 13. Comparison of the relative temperatures in crystals grown in various
media and shielding modes (the straight line represents the temperature in
crystals grown in vacuum with standard shielding.

In the course of analyzing the experimental data obtained under various shielding conditions and in various media it is instructive to examine the deviations of the calculated temperature in the crystals from the straight line obtained for vacuum without special shielding. For this purpose we plotted on the graph showing this line (Fig. 13) the relative temperatures in the crystals, calculated from the given formula for typical melting operations. The scatter of the points relative to the straight line is indicative of the particular shielding system and medium and their effect on the temperature field of the crystal.

Growing of Crystals in Vacuum with a Floating Shield

Graphite shields of various configurations, floating on the surface of the melt, including those shown in Fig. 3, were investigated in the industrial apparatus (Figs. 1, 2, 4) in a high vacuum of 0.0133 N/m². The diameter of the base of the shield was 2 or 3 mm smaller than the inside diameter of the cylindrical crucible, with three or four pins around the periphery of the base centering the shield with respect to the axis of the ingot.

The temperature of the shield wall parallel to the surface of the melt (Fig. 3, No. 1) was 850°C, diminishing gradually in the conical section of the shield to 770-730°C at the top. The wall temperature of the conical shield (Fig. 3, No. 2) varied from 900°C at the base of the cone to 770-730°C at its top. The mean wall temperature of the overhead shield used for standard shielding amounted to 700-600°C, depending on its distance from the level of the melt. The investigated shields differed from one another in height (40-60 mm) and diameter of the opening for emergence of the ingot (50-80 mm). The thickness of the shield walls was 2 or 3 mm. The angular shielding coefficient φ of the shields was roughly the same.

The experiments showed that with a ratio of the shield opening diameter to crystal diameter $d_{sh}/d_{cr} \geq 3$ the floating shield does not alter the temperature gradients in the ingot relative to growth in vacuum with the standard shielding arrangement. This is attributable to the large angle subtended by the surface of the ingot in the shielded zone relative to the cold walls of the enclosure. The cooling rate m in this case is the same in the shielded zone as in the unshielded zone.

For $d_{sh}/d_{cr} < 3$ the shielding of the ingot becomes effective. In the zone of the floating shield (h_{sh} = 40–60 mm above the level of the melt) the cooling rate of the ingot m is lower than in the unshielded zone (the values of m for melts in vacuum without a floating shield are shown in the table for comparison, tests Nos. 1–11). In the graph of $\ln t/t_0$ vs. Fo (Fig. 9) the lines suffer a bend at values of Fo corresponding to segregation of the ingot from the shielded zone. The floating shield with $d_{sh}/d_{cr} < 3$ tends to decrease the radial gradients in the shielded zone by slowing down the cooling rate of the ingot in vacuum.

On emerging from the shields, the cooling rate of the ingot is determined by radiative heat transfer with the water-cooled walls of the enclosure. The cooling rate in this case is increased, which leads to an increase in the radial gradients.

The graph (see Fig. 13) is marked by a rather characteristic heating of the ingot in the shielded zone and more rapid cooling (beginning with t = 750°C) on emerging from the shielded zone.

The experiments with a floating shield disclosed the interesting fact that the cooling rate of the ingot m is independent of the shield geometry. The results of processing the data in coordinates $\ln t/t_0$ vs. Fo, obtained with the standard shielding arrangement, are shown in Fig. 9 for a shield with an opening 60 mm in diameter and for two different types of floating shield (Nos. 1 and 2). It is apparent that the cooling rates for slightly different shapes and pulling rates of the ingots in the shielded zone were the same in all three cases.

Clearly, given purely radiative heat transfer as occurs in a highly rarefied atmosphere (vacuum of 0.0133 N/m²), with a high-temperature radiation source near the crystal (temperature of the melt surface ≈ 940°C), changing the configuration of the shields does not significantly alter the temperature field of the ingot.

A simple estimate of the normalized emissivity for a sealed enclosure according to the relation

$$\varepsilon_N = \frac{1}{\dfrac{1}{\varepsilon_{cr}} + \dfrac{F_{cr}}{F_{sh}}\left(\dfrac{1}{\varepsilon_{sh}} - 1\right)}$$

shows that changing the area ratio of the shielding surface to the crystal surface from $F_{sh}/F_{cr} = 5$ to $F_{sh}/F_{cr} = 2$ changes the value of ε_N 3.0 to 7.0% relative to ε_{cr}. This gives rise to a 3.0–7.0% change in the radiant heat flux Q_r from the crystal, given identical shield temperatures:

$$Q_r = \varepsilon_N \sigma_0 F_{cr}\left[\left(\frac{T_{cr}}{100}\right)^4 - \left(\frac{T_{sh}}{100}\right)^4\right]$$

whereas the change in temperature of the crystal, being proportional to $\sqrt[4]{Q_r}$, comes to only 1.5–2% (which lies within the experimental error of the temperature measurements).

The data obtained under industrial conditions on the quality of the single crystals show that, for growth in high vacuum, floating shields do not exert any appreciable influence on the dislocation density in the crystals. This, without a doubt, is related to the incapacity of such shields to significantly alter the temperature field of the crystal near the crystallization front.

Growing of Crystals in Vacuum with an Overhead Shield

The temperature fields of crystals grown in a vacuum of 0.0133 N/m² using stationary graphite shields, isolating the entire surface of the crystal from the water-cooled walls of the enclosure (Fig. 2), were investigated in the industrial apparatus. The use of this type of shield is aimed at decreasing the thermal stresses in the crystal, which tends to exchange heat with the cold walls of the apparatus. The mean temperature of the two cylindrical shields was ∼ 750°C at the level of the melt, about 450°C in the upper region (the average temperature of the inner wall of the enclosure over its height did not exceed 80–100°C for all the pieces of apparatus used).

The results of calculating the crystal temperature in one melting operation by the semiempirical formula is shown in Fig. 13. There is a noticeable systematic deviation in the resultant data (beginning with t = 720–750°C) from the temperature of a crystal grown without special shielding.

However, only in the temperature range below 500°C does the given type of shielding significantly diminish the radial gradients in the crystal (20 to 25%) by comparison with pulling in vacuum without special shielding; in this range a change in the heat exchange conditions cannot very well produce plastic deformations in the crystal, thereby altering its structure. A comparison with statistical data on the quality of single crystals confirms these results. The presence of overhead shields does not change the dislocation density as compared with standard shielding and cannot affect the impurity distribution in the crystal, since no changes are introduced in the temperature field of the ingot near the crystallization front. It is clear in Fig. 13 that the behavior of the temperature in the crystal in the interval below 800°C coincides with the behavior with standard shielding.

Growing of Crystals in an Inert Gas Atmosphere

The presence of a gas can substantially alter the heat transfer conditions in the enclosure and affect the temperature field of both the crystal and the melt.

The theoretical calculation of the composite heat transfer under conditions of free convection in a bounded enclosure in the presence of a heat source with nonstationary temperature variation is exceedingly difficult, even without shielding of the crystal or melt, and requires the setting up of special experiments aimed at determining the convective heat transfer coefficient. On the other hand, the pulling process in a gaseous medium is of unquestionable interest to industry, as this obviates the need for creating vacuum in the system.

The growing of crystals in a gas atmosphere does not deteriorate the quality of the single crystals as compared with growth in vacuum.

In order to estimate the influence of the medium, we conducted experiments to ascertain the temperature fields of crystals grown from the melt in an inert gas atmosphere.

Due to the considerable chemical activity of germanium at high temperatures, the seed-pulling process can only be effected in an atmosphere of inert gases (Ar, He) or some kind of shielding atmosphere (for example, a hydrogen atmosphere). Argon is the most important medium for industry from the viewpoint of ease of purification. The process is run in the absence of streaming at a pressure differential of about $0.294 \cdot 10^5$ $N/m^2 = 0.4$ atm.

The investigations showed that an argon atmosphere with standard shielding slightly affects the temperature field of the crystal. A certain cooling of the crystal in the zone affected by the overhead shield (up to t = 650°C) and heating of the crystal in the unshielded zone are evident in Fig. 13.

Argon, helium, and hydrogen are diathermanous, i.e., essentially transparent to thermal radiation. Consequently, the change in temperature field of the crystal from that in vacuum, as shown in Fig. 13, is related solely to the additional transfer of heat due to natural convection from the hot regions of the enclosure (surface of the melt) to the cooler regions (crystal and wall of the enclosure in the unshielded zone). The presence of convective heat flux during melting in argon lowers the cooling rate of the ingot relative to growth in vacuum with the same ingot geometry and pulling parameters (Fig. 10). For growth in an argon atmosphere only a slight increase is observed in the radial gradients in the crystal in the shielded zone, with a decrease in the unshielded zone (but no more than 5 or 6%). The weak effect of argon is explained by its very low thermal conductivity (λ_{Ar} ranges from 0.0349 to 0.0582 W/m°C in the temperature interval from 500 to 1000°C at atmospheric pressure).

Consequently, the presence of argon in the enclosure with P_{Ar} = 0.3 atm cannot produce significant alterations in the structure of the crystal. This is confirmed by statistical data on the quality of single crystals.

It seemed appropriate to investigate the effect of various pressures of the medium on the pulling process. In this connection experiments were conducted in the apparatus (see Fig. 4) to determine the temperature fields of the crystals and melt with argon pressure differentials from $0.098 \cdot 10^5$ to $1.17 \cdot 10^5$ N/m² (0.1 to 1.2 atm), standard shielding, and an open melt surface. Although the results of the data processing are not given here, the following deserves mention.

A change in pressure P_{Ar} from 0.098 to $1.17 \cdot 10^5$ N/m² did not cause any significant change in the temperature field of the crystal as compared with the vacuum process. With the absence of overhead shielding considerable cooling of the melt surface was observed, which led to the onset of extra crystallization centers and upset the seed pulling process. Cooling of the melt surface in the absence of shielding is related to the intense convective transfer of heat to the cooler region of the enclosure. The crystals grown under these conditions were characterized by an extremely irregular resistivity in the cross section of the ingot and high dislocation density ($N_d \geq 10^4$ cm⁻²).

The presence of an overhead shield inhibited free convection from the surface of the melt. An analysis of the temperature field of the crystal disclosed that in this case the stream of hot gases rising from the surface of the melt through the opening in the shield heats the crystal, promoting a certain reduction of the radial gradients on its surface (by 4 or 5%) in the temperature interval 850–650°C. Despite the insignificant variation in temperature field of the crystal by comparison with the vacuum process, an appreciable improvement was observed in the quality of the single crystals for $P_{Ar} = 0.588 \cdot 10^5$ N/m² (uniform resistivity over the end face, $N_d = 10^2$ to 10^3 cm⁻²), which clearly was induced by the influence of the temperature field of the melt. In fact, additional measurements of the temperature field of the melt at elevated pressures of the medium disclosed stable isotherms beneath the crystallization front during growth of the ingot, which was not observed when the surface of the melt was exposed.

Growing of Crystals in Argon with Various Shielding Modes

Investigations were conducted in the industrial apparatus on the temperature fields of crystals grown in an argon atmosphere ($p = 0.294 \cdot 10^5$ N/m²) with shielding of the entire surface of the ingot (overhead shield, Fig. 2) and with various types of floating shield.

The investigations showed that the overhead shield has the same effect on the ingot as in vacuum (heating of the ingot in the low-temperature range is evident in Fig. 13, beginning with 620°C), although the presence of convection currents associated with the presence of argon in the enclosure somewhat diminishes the effect of this heating. The quality of the single crystals, as already stated, is not significantly improved by such shielding.

The occurrence of a convective heat transfer component with growing in an inert gas medium alters the distribution of the temperature field of the crystal by comparison with its growth in vacuum and in the presence of floating shields. Heating the shield, the argon carries heat from the surface of the melt and the ingot, causing a certain increase in the radial gradients on the surface of the crystal in the shielded zone. In the unshielded zone, on the other hand, the crystal is heated by streams of hot gases rising through the opening in the shield. In the graph of Fig. 11, which shows the processing of the thermograms in the coordinates $\ln t/t_0$ vs. Fo, a characteristic change in the cooling rate m is clearly seen. This change sets in after the crystal emerges from the shielded zone. This is related to convective heat transfer; the gas stream flows past the crystal above the opening in the shield, continuing to remove heat from its surface (until t = 750–650°C, depending on the height of the floating shield), transporting it to the colder regions of the crystal and walls of the enclosure. In the presence of a gaseous medium the configuration of the floating shields exerts an influence on the process of heat exchange between the crystal and the surrounding medium.

The slight change in the temperature field of the crystal when grown in argon with floating shields cannot significantly affect the quality of the single crystals. At the same time, a qualitative analysis of more than 40 ingots grown with floating shields and without them has shown that pulling from a melt protected by a floating shield yields a somewhat lower dislocation density ($N_d = 10^2$ to 10^3 cm^{-2}). This is clearly related to the favorable influence of the floating shield in the presence of argon on the thermal processes occurring in the melt and at the crystallization front.

Growing of Crystals with Extremal Temperature Gradients

The investigations of the temperature fields of crystals grown under the conditions described above have shown that the shielding modes employed in industry have too little effect on the temperature field of the single crystal and cannot produce a significant improvement in quality of the crystal.

In a number of studies relating to this problem, however, it is indicated that keeping germanium specimens for a long period of time at high temperatures makes it possible to arrive at a redistribution and even a diminution of the dislocation density. In this connection, we have begun experiments in the laboratory apparatus, growing germanium single crystals under the conditions present with minimum and maximum temperature gradients in the crystal.

A theoretical evaluation of shielding surfaces shows that the heat flux from the surface can be minimized with a shield radius just slightly smaller than the largest radius of the crystal. We therefore grew a crystal under the following conditions. Over the melt we suspended a graphite shield (Fig. 7), which was heated by a current at standard frequency. The inside diameter of the shield was chosen with the provision that capillary tubes containing thermocouples could be inserted into the shield for measurement of the temperatures in the ingot (D_{sh} = 36 mm with ingot diameters of 20 to 23 mm). During the pulling process the crystal was slowly displaced in the shield and after separation from the melt was gradually cooled in high vacuum to a temperature $t \approx 150°C$. At a shield temperature of 850°C a reduction in the radial gradients was achieved by comparison with growth of the crystal under the same conditions without shielding, on the average 30% less over the length of the ingot.

A qualitative analysis of the grown single crystals showed that the dislocation density (after segregation at a constant diameter of 20 to 23 mm) remained almost unchanged over the length of the ingot, amounting to $2 \cdot 10^3$ to $5 \cdot 10^3$ cm^{-2}. The etch pits in this case were uniformly distribution in the cross section of the sample. After growing, the crystal was cooled in a heated shield with a smooth drop in temperature. With sudden drops in shield temperature an increase was observed in the dislocation density in a surface layer of the crystal approximately 2 mm thick. In spite of the small radial gradients in the crystal, anodic etching of a lengthwise section of the ingot did not disclose a uniform impurity distribution (in general, a thin channel with low resistivity ran along the ingot). This fact indicates that the impurity distribution in the crystal is apparently related to the configuration of the isotherms in the melt and is determined by the thermal conditions at the crystallization front. The heated shield, by tending to equalize the temperature field of the ingot at a distance of 20 mm from the melt, decreases the number of defects in the crystal lattice which are associated with thermal stresses in the crystal (mainly the number of dislocations) but cannot significantly influence the impurity distribution in the solidified single crystal.

The heated shield reduces the thermal stresses in the crystal, to a considerable extent inhibiting the possibility of added nucleation of defects in the structure of the ingot at the plasticity temperatures of germanium. The use of such shields (or shields of material with a very low coefficient of thermal conductivity, for instance quartz glass) with a diameter near the diameter of the ingot is very suitable under industrial conditions.

For the purpose of investigating the effect of maximum temperature gradients on the quality of single crystals, we performed several experiments in the laboratory apparatus with a water-cooled ring situated at various distances from the crystallization front. Lowering the "cold" ring below 10 mm from the surface of the melt caused crystallization of the melt beneath the ring and upset the pulling process.

Systematic measurements of the temperature fields of the crystals were not made under these conditions, but a qualitative analysis indicated a high dislocation density ($N_d = 10^4$ to 10^5 cm^{-2}), the presence of dislocation lines in the crystal cross sections, and a large scatter in values of the resistivity at the ends of the ingot. Clearly, it is unsuitable to maintain maximum temperature gradients in the ingot near the crystallization front.

Growing of Crystals with a Small Number of Dislocations

Dislocation-free germanium crystals were prepared under production conditions with the industrial apparatus (see Fig. 4) in an argon-hydrogen atmosphere with a definite shielding of the crystal and melt surfaces. It was of decided interest to study the temperature field of the crystal under these conditions and to estimate the influence of the given type of shielding.

Experiments to measure the temperature of the crystal during its growth were performed under factory conditions in the industrial apparatus. The surfaces of the melt and crystal were shielded by a floating graphite shield with a quartz sleeve (Fig. 5). The melting was carried out in an argon and hydrogen atmosphere with the mixture at a pressure $p_{mix.} = 0.343 \cdot 10^5$ N/m^2. The crystals were grown with a diameter of 25 to 28 mm and a length of 180 to 250 mm, with a minimum value of $N_d = 10$ cm^{-2}. The pulling rate in the experiments varied from 2 to 1.6 mm/min, the crucible and seed were rotated at 8 and 35 rpm, respectively. Instantaneous cessation of rotation of the seed (with the implantation of thermocouples) resulted in an increased dislocation density, to $N_d = 1 \cdot 10^2$ cm^{-2}. For this reason the ingot was usually pulled with the seed crystal still. Stopping of the crucible also gave rise to an increase in dislocation density and nonuniform impurity distribution in the cross section of the ingot, which is associated with displacement of the floating shield and violation of the symmetry of the temperature field of the melt when the crucible is stopped. In the experimental melting operations the rotation of the crucible was maintained at a constant rate.

A comparison of the data from the temperature measurements in crystals grown under these conditions with the temperature field of a crystal in vacuum produced an unexpected result. The change in temperature inside the crystal scarcely differed from the case of pulling in vacuum without special shielding. This is explained by the additional influence of the quartz sleeve shielding the ingot at a distance of 12 mm from the surface of the melt. Because it is invested with a very small thermal conductivity (approximately 1/50 the thermal conductivity of graphite at t = 800°C), the quartz shield contains heat about the crystal at the shielding height h_{sh} = 35–40 mm, thus compensating for the convective transfer of heat by the gaseous medium to the cooler part of the enclosure.

It now becomes understandable why the pulling rate exerts so considerable an influence on the temperature field of the ingot as observed with this shielding system. Figure 13 shows the deviation in temperature in the ingot with the given type of shielding in comparison with cooling of the ingot in vacuum without special shielding. At a pulling rate of 2 mm/min the variations in temperature under the experimental conditions and in vacuum coincide; with the pulling rate decreased to 1.6 mm/min the temperature in the crystal is lowered.

This happens because when the pulling rate is decreased the heat of crystallization developed per unit time is lowered. Under the thermal conditions created by a gaseous medium and the quartz shield the heat of crystallization begins to have a considerable influence on the thermal effects near the crystallization front. This also explains the fair increase in radial gradient in the crystal in the shielded zone when the pulling rate is decreased from 2 to 1.6 mm/min.

In order to corroborate the effect described above in connection with the influence of the quartz sleeve, we conducted experiments under similar conditions, but with a graphite sleeve (see Fig. 5b). As expected, graphite, in that it has a fairly high thermal conductivity, is far less effective as a shield for the crystal. The radial gradient in the crystal increases approximately 6–7% by comparison with growth under the same conditions with a quartz sleeve, in the range of crystal temperatures equal to 900–750°C (shielding zone).

The fact that the temperature field of the crystal in the environment used to grow dislocation-free single crystals is almost the same in the growth of crystals in vacuum without special shielding (when it is impossible to obtain $N_d < 10^2$ cm^{-2}) leads to the conclusion that structural defects originate at the crystallization front, and sudden fluctuations of temperature during cooling of the crystal in the enclosure can quite clearly aggravate them.

The following conclusions are based on the experimental investigations that we conducted:

1. It was found that with standard shielding in vacuum as applied in furnaces (Figs. 1 and 2) the shape of auxiliary floating shields does not affect the temperature field of the crystal. From the point of view of reducing the thermal stresses in the ingot in the shielding zone the choice of floating shields of various configurations is inappropriate. It is much more effective to shield the surface of the melt, creating a symmetrical temperature field therein and maintaining a constant temperature gradient at the crystal-melt interface. The latter, however, must be confirmed by careful investigation of the influence of the temperature field of the melt and the processes taking place at the crystallization front on the formation of a highly perfect single crystal structure.

2. The greatest reduction in radial gradients on the surface of the ingot was obtained in the case when overhead shields were used in vacuum, isolating the surface of the crystal from the cold walls of the enclosure. However, a significant reduction in radial gradient is observed only for t < 550°C, where germanium is essentially nonplastic and a reduction in thermal stresses in the crystal cannot do anything to improve its structure.

3. All investigated media and shields (except those described in the growing of crystals with extremal temperature gradients and a small number of dislocations) used under factory conditions yielded at best 10% changes in the temperature gradients in the ingot in the range 850 to 650°C by comparison with vacuum and the standard shielding arrangement. This could not significantly lower the thermal stresses in the crystals or improve their quality, a fact confirmed by qualitative analysis of the grown ingots.

4. It is convenient for the reduction of thermal stresses in the crystal to use a warmed shield whose diameter approximates that of the ingot.

5. The results obtained in our investigation of the temperature field of an ingot under conditions appropriate to the growth of dislocation-free crystals (absence of any variation in temperatures of the ingot above the crystallization front by comparison with pulling in vacuum without special shielding) provide still another argument in favor of the assumption that the presence of defects in the structure of a single crystal is governed primarily by the thermal conditions at the crystallization front.

Literature Cited

1.   F. D. Rosi, R.C.A. Review, 19 (3):349 (1958).
2.   E. Billig, Proc. Roy. Soc., Series A, 235:37 (1956).
3.   M. Francois, Solid State Physics in Electronics and Telecommunications, Vol. 1: Semiconductors, p. 171 (1960).
4.   P. Pennig, Philips Res. Reports, 13:79 (1958).
5.   D. Hahn and D. Taube, Z. Angew. Phys., 15:2 (1963).
6.   N. B. Vargaftik (editor), Thermophysical Properties of Substances (Gosénergoizdat, 1956).
7.   C. A. Slack and C. Glassbrenner, Phys. Rev., 120:782 (1960).

EXPERIMENTAL DETERMINATION OF THE EMISSIVE POWER OF GERMANIUM AND SILICON IN THE TEMPERATURE RANGE 700-1200°K

V. F. Brekhovskikh

In carrying out the engineering calculations associated with the thermal conditions in equipment for the preparation of single crystals of semiconductor materials, it is necessary to know the total emissivity of the latter over a wide range of temperatures, particularly in the vicinity of the melting point. The literature contains data on the optical properties of germanium and silicon, but these have only been determined for isolated regions of the spectrum at relatively low temperatures (up to 600°C). However, because the spectral and total emissivities may differ considerably, it is important to investigate experimentally the total and spectral emissivities of germanium and silicon. In so doing, it must be borne in mind that when germanium is heated in air to temperatures of the order 1000°K one begins to encounter the intense sublimation of germanium monoxide formed on the surface of the sample. Consequently, the investigation of its optical properties at high temperatures must be carried out in a vacuum or in an inert gas atmosphere.

The apparatus used to determine the emissivity (Fig. 1) is a water-cooled steel chamber 1 mounted on a flange 2, which is also cooled by water. The flange has openings for electrical leads 3 and thermocouples 4, as well as a tubing to connect the chamber with the vacuum pumps. The inner surfaces of the chamber and flange are blackened with a mixture consisting of lamp black and BF-4 cement. The emissivity of this mixture is approximately equal to 0.96. A thermopile 5 with light guide 6 is fitted to the chamber housing. The thermopile is made up of ten Chromel-Alumel thermocouples fabricated of 0.07 mm wire and installed in a heavy metal casing with the inside surface blackened. The temperature of the latter (hence the temperature of the cold thermocouple junctions) is held constant by means of a thermostat. The working ends of the thermocouples are bonded in thin plates and blackened. The radiant energy flux from the hot sample is transmitted to the sensitive surface of the thermopile through the light guide, which is a cylindrical rod of synthetic corundum 3.5 mm in diameter and 200 mm long. The light guide is rigidly secured to the thermopile casing by means of a special attachment. Its midsection passes through a water-cooled collar 9.

The application of a light guide enhances severalfold the sensitivity of the thermopile, in that the energy losses in it are small. This is because almost all light rays passing into the light guide through its end face are propagated along it due to total internal reflection from the lateral surface, which is highly polished. Light guides are ordinarily made of synthetic corundum, which combines excellent optical and mechanical properties.

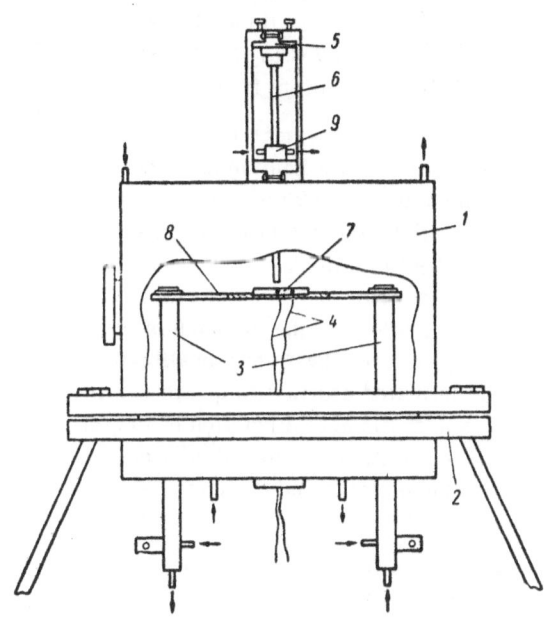

Fig. 1. Diagram of the apparatus used to determine the emissivity.

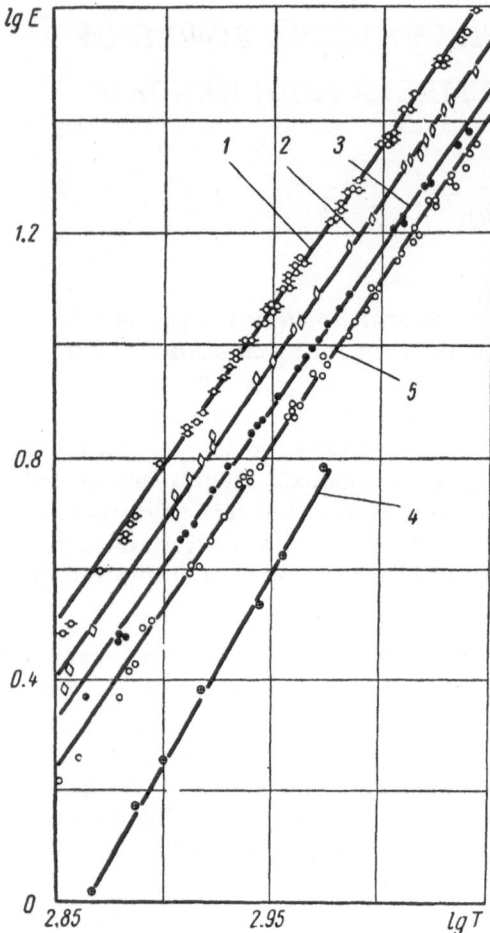

Fig. 2. Dependence of the thermopile signal amplitude on absolute temperature. 1) Black body model; 2) graphite; 3) mechanically polished germanium; 4) strongly oxidized aluminum; 5) dip-polished germanium.

In particular, the transmittance of corundum in the infrared is considerably in excess of the transmittance of optical quartz. Moreover, the thermopile light guide can be aligned on a small section of the surface of the test sample.

The thermopile and light guide were calibrated with black body radiation. The black body model was a cavity formed by two closely fitted graphite heaters, each with a recess 2 mm deep and 27 mm in diameter. The thermopile was aimed at a hole 4.5 mm in diameter in the top of the heater. The wall temperature of the cavity was measured with Chromel–Alumel thermocouples inserted in this cavity. Consequently, all of the requirements for a black body model were established; the area of the opening was sufficiently small in comparison with the total surface area of the cavity, the temperature was uniform at all points of the cavity. A numerical calculation yielded a value of $a = 0.995$ for the absorptive power of the model. Making allowance for the temperature measurement error, the value may be assumed equal to 0.98.

The quality of the black body model may be judged from a graph showing the dependence of the thermopile signal amplitude on the absolute temperature, in log–log coordinates (Fig. 2). As apparent from the graph, the experimental data lie well along a straight line in the interval from 700 to 1100°K. At temperatures above 1100°K slight deviations from linearity are observed.

The following experimental procedure was used. The sample (Fig. 1), a germanium or silicon wafer 2.5 to 3.5 mm thick and 18 to 20 mm in diameter, was placed in the well of the heater 8. The heater was a graphite plate 2–3 mm thick and 27 mm wide, mounted on electrical conductors 3. Two holes were drilled into the sample from below for the insertion of one or two Chromel–Alumel thermocouples with a thickness of 0.2 mm. The thermocouples were precalibrated against a platinum–platinorhodium thermocouple. The thermocouples were insulated either with quartz capillaries or by means of a refractory compound, but in either case the distance from the active junction of the thermocouple to the surface of the sample was never more than 0.5 to 1.0 mm. As shown by calculations, the total temperature measurement error due to heat loss through the thermocouple leads and the presence of a temperature drop across the height of the sample did not exceed 1%.

Above the heater was the light guide 6, along which radiant energy flux was transmitted from the heated sample to the sensitive surface of the thermopile. The lower end of the guide was separated from the sample surface by 1.5 to 2.0 mm. As shown by tests with graphite, the thermopile signal amplitude at a given temperature did not vary (within the limits of experimental error) with changes in the distance between the light guide and heater from 1.0 to 5.0 mm over the entire range of investigated temperatures. The signal from the thermopile and from the thermocouples was delivered to the input of a high-resistance dc potentiometer. The experiments were carried out in vacuum at pressures of the order of 0.0532 N/m² (except for the experiments with certain metals, when the pressure in the chamber was about 3.99 N/m²).

The emissivity of the investigated material was determined from the relation

$$\varepsilon = 0.98 \frac{E}{E_0},$$ (1)

where E is the thermopile signal amplitude in the case of the sample, mV; E_0 is the thermopile signal amplitude in the case of the black body model, mV.

The numerical coefficient 0.98 represents the absorptive power of the black body model.

In our work we investigated germanium and silicon samples with various grades of surface finish (chemical and mechanical polishing). The germanium was chemically polished with a dip solution of 50% HNO_3, heated to boiling. The silicon samples were dip-polished in a mixture of 30% HF and 70% HNO_3. A grinding powder with a grain size not exceeding 10 μ and boron carbide abrasive powder were used for the mechanical polishing. Also investigated was a germanium sample cut from the single crystal such that its natural surface was retained with the growth bands.

In order to ascertain the laws governing emission from germanium and silicon, a dependence of the form log E = f(log T) was constructed. It is evident in Fig. 2 that in the case of mechanically polished germanium a straight line is obtained, its slope equal to the slope of the calibration line constructed for the black body model. It may be concluded, therefore, that its emissive power is proportional to the fourth power of the absolute temperature. For dip-polished germanium a slight deviation from the fourth-power law is noted in the temperature interval from 700 to 1100°K. The same effect is observed in the case of the silicon, except for the difference that here there is no divergence between the straight lines constructed for the mechanically and chemically polished samples. The same graph shows data obtained for graphite and aluminum. It follows from a comparison of the lines for the black body model and for aluminum that the emissive power of the latter is proportional to $T^{4.7}$ (the quantity indicated in [1] is $T^{4.81}$).

The dependence of the emissivities of germanium and silicon on the temperature and grade of surface finish in the interval 700 to 1200°K is shown in Fig. 3. For silicon, which began to oxidize under the experi-

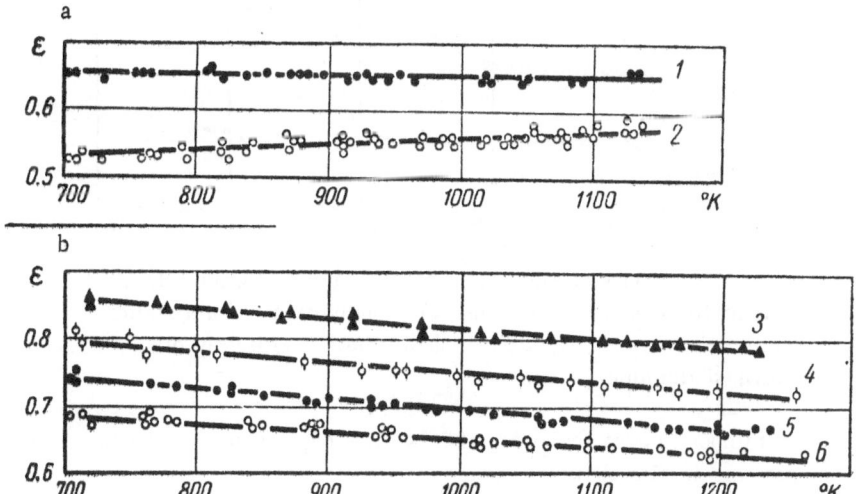

Fig. 3. Dependence of the total emissivity on temperature. a) Germanium: 1) mechanically polished; 2) dip-polished; b) silicon: 3) strongly oxidized (heating at 1270°K for 10 min); 4) mildly oxidized (heating at 1270°K for 5 min); 5) mechanically polished; 6) dip-polished.

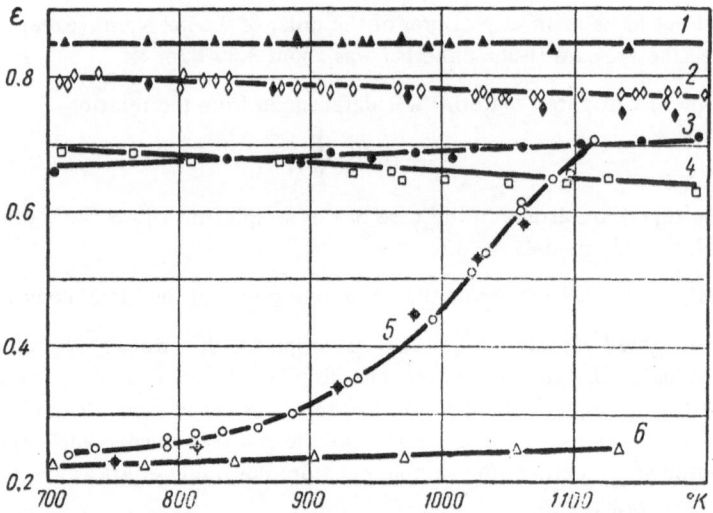

Fig. 4. Emissivity of graphite and steel. 1) Oxidized steel; 2) graphite; 3) stainless steel after rolling and heat treatment (heating at 1120°K for 15 min); 4) mechanically polished stainless steel; 5) stainless steel after rolling; 6) steel with a lightly dip-polished surface.

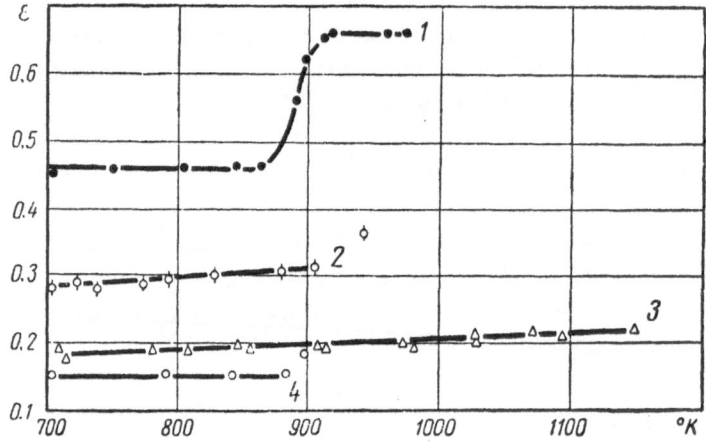

Fig. 5. Emissivity of aluminum, nickel, and Duralumin. 1) Mechanically polished Duralumin; 2) strongly oxidized aluminum; 3) nickel with a lightly dip-polished surface; 4) lightly dip-polished aluminum.

mental conditions at a temperature of the order 1270°K, the dependence of the emissivity on the keeping time at this temperature is also shown.

To confirm the validity of the measurement procedure used with the given apparatus, we determined the emissivity of materials that had already been investigated. We first studied the heater material, graphite, obtained by sintering of a mixture consisting of pitch, soot, and a binding agent. The variation in its emissivity with temperature is shown in Fig. 4. Data borrowed from [2] are shown in the same figure. It is apparent from the graph that a slight departure is observed only in the temperature interval from 1100 to 1200°K. Good agreement is also obtained for stainless steel after rolling [3]. The temperature dependence of the emissivity for aluminum, nickel, and Duralumin is shown in Fig. 5. A sample of strongly oxidized aluminum (its emissivity equal to 0.29 at 775°K, in agreement with the tabulated data [1]) was melted at a temperature of the order 940°K, as a result of which its emissivity was markedly increased. The increased emissivity of Duralumin at a temperature of about 890°K is explained by oxidation of its surface. The results obtained for nickel are difficult to compare with existing data, because a sample of heavily contaminated metal was used for the experiment.

In this arrangement we determined the emissivity of gallium antimonide and molten germanium. The surface of the GaSb sample was chemically polished. Its emissivity turned out to be equal to 0.58−0.60 in the temperature interval 700 to 850°K. Further heating resulted in the intense liberation of antimony, which became deposited on the light guide.

The germanium was placed in a quartz cup 20 mm in diameter and 15 mm high, which contained branch tubes for the thermocouples. Otherwise the experiment was conducted as before. The experiments showed that when germanium goes from the solid to the liquid phase its emissive power falls off sharply. Whereas the emissivity of solid germanium near the melting point is 0.58, for liquid germanium it comes to no more than 0.18. This is apparently related to the change in structure of germanium and its electrophysical properties in transition to the liquid phase [4]. Solid germanium crystallizes in a diamond type lattice, while in liquid germanium the atoms are more densely packed. This attests to the fact that the binding forces in liquid germanium have a more metallic nature. This is further evinced by the fact that the resistivity of liquid germanium at the melting point is just 1/15 the resistivity of solid germanium.

Knowing the emissivity of liquid germanium, we can attempt to estimate its thermal conductivity. To do this we make use of the heat balance equation at the free surface of the melt:

$$\lambda_l \frac{\partial t}{\partial z} = \varphi_{12}\varepsilon_N\sigma_0\left(t^4 - t^4_w\right), \tag{2}$$

where λ_l is the thermal conductivity of liquid germanium, $\partial t/\partial z$ is the axial temperature gradient in the melt near its surface, φ_{12} is an angular coefficient of the system (melt surface + chamber wall of the apparatus) and is equal to 0.5, σ_0 is the Stefan−Boltzmann constant, t is the temperature at the surface of the melt, t_w is the wall temperature of the apparatus ($t_w \ll t$), ε_N is the normalized emissivity, defined according to the formula

$$\varepsilon_N = \frac{1}{1 + \varphi_{12}\left(\frac{1}{\varepsilon_l} - 1\right) + \varphi_{21}\left(\frac{1}{\varepsilon_w} - 1\right)},$$

where ε_l is the emissivity of liquid germanium, ε_w is the emissivity of the chamber walls (approximately equal to 0.6), φ_{21} is an angular coefficient of the system (chamber walls + melt surface) (since the quantity is small, the last term in the denominator above may be neglected.

The investigations were carried out in the apparatus by the Czochralski seed-pulling technique. It is important to note that convection plays a large part in the heat transfer in molten germanium, a fact which greatly complicates measurement of the temperature gradients. However, near the surface of the melt there exists a boundary layer in which the heat transfer process occurs solely due to heat conduction. The axial temperature gradient near the surface of the melt was determined by means of two Chromel−Alumel thermocouples placed in quartz sheaths and connected in a differential circuit. The upper thermocouple was situated directly at the surface of the melt, the lower one at a vertical distance of 8 mm. The quartz sheaths were bent so that the length of the section of thermocouple immersed in the melt was at least 25 mm, which meant that the measurement error due to heat loss in the thermocouple leads was minimized. We necessarily point out that the magnitude of the gradient depends on whether the crucible and the melt contained therein are rotated or not. Without rotation $(\partial t/\partial z)_1 = 250$ deg/m, with rotation at about 7 rpm $(\partial t/\partial z)_2 = 500$ deg/m (at a temperature t = 1215°K). The gradient did not change when the speed of rotation of the crucible was stepped up. Substituting the indicated values into Eq. (2) and recognizing that $\varepsilon_N = 0.307$, we obtain $\lambda_{l_1} = 77$ W/m · deg and $\lambda_{l_2} = 38$ W/m · deg. The first value is clearly too high, but it gives some idea as to the upper limit of the thermal conductivity of liquid germanium near the melting point.

Literature Cited

1.   A. G. Blokh, Fundamentals of Radiative Heat Transfer (Gosénergoizdat, 1962).
2.   Collection: High-Temperature Research [Russian translations] (IL, 1962).
3.   S. De Corso and R. Coit, Trans. ASME, 77 (8):(1955).
4.   R. W. Keyes, Phys. Rev., 84 (2):(1951).

INVESTIGATIONS OF THE TEMPERATURE FIELD OF A GERMANIUM MELT IN THE GROWING OF SINGLE CRYSTALS BY THE CZOCHRALSKI METHOD

V. N. Chakhunashvili

The Czochralski method enjoys widespread application for the production of germanium and silicon single crystals. The growing process is conducted in an inert or reducing medium, sometimes in vacuum. The diameter of the crystal is regulated by varying the pulling rate or the power input. Both means are sometimes employed. To impart symmetry to the temperature field of the melt, the crystal is rotated during its growth. At the interface between the liquid and solid phases, latent heat of crystallization is continuously developed and must be eliminated. For the growth process to be stable an appropriate temperature gradient must be mounted on the axis of the crystal. Usually in the melt a temperature gradient is created such that the temperature increases with distance from the interface. In this case there occurs a certain net heat flux from the melt to the crystal through the interface.

Experimental investigations carried out in [1] show that the crystallization process is roughly determined by the following factors:

    a. Crystallographic characteristics (crystal symmetry and nature of the binding forces;

    b. Temperature distribution in the melt and solid metal;

    c. Impurity distribution in the melt.

The laboratory for thermal operation of vacuum systems at the Moscow Institute of Railroad Engineers (MIIT) has conducted a research effort directed at the temperature fields of crystals and their modes of shielding during growth.

The crystal pulling process is significantly affected by the form of the solid–liquid interface.

The macroscopic structure of the interface is primarily determined by the rate of heat emission from it and the rate of diffusion of dissolved impurities. The interface surface may take the shape of an isotherm or some more complex form, depending on the redistribution of impurities near the interface, as well as the nature of the heat transfer. The form of the interface depends on the magnitude of the surface forces, temperature gradient, crystal growth rate, mean impurity concentration, impurity distribution coefficient, and diffusion coefficient of this impurity in the melt. For this reason, therefore, considerable importance attaches to the investigation of the temperature field in the melt.

The thermal process involved in the melt during the growth of single crystals is described by the equation of continuity:

$$\frac{\partial w_z}{\partial z} + \frac{\partial w_r}{\partial r} + \frac{w_r}{r} = 0, \tag{1}$$

by two equations of motion:

$$\frac{\partial w_r}{\partial \tau} + w_r \frac{\partial w_r}{\partial r} + w_z \frac{\partial w_r}{\partial z} = g_r - \frac{1}{\rho} \frac{\partial p}{\partial r} + \nu \left(\frac{\partial^2 w_r}{\partial r^2} + \frac{\partial^2 w_r}{\partial z^2} + \frac{1}{r} \frac{\partial w_r}{\partial r} - \frac{w_r}{r^2} \right), \tag{2}$$

$$\frac{\partial w_z}{\partial \tau} + w_r \frac{\partial w_z}{\partial r} + w_z \frac{\partial w_z}{\partial z} = g_z - \frac{1}{\rho} \frac{\partial p}{\partial z} + \nu \left(\frac{\partial^2 w_z}{\partial r^2} + \frac{\partial^2 w_z}{\partial z^2} + \frac{1}{r} \frac{\partial w_z}{\partial r} \right), \tag{3}$$

and by the equation for heat transfer in the melt:

$$\frac{\partial t}{\partial \tau} + w_r \frac{\partial t}{\partial r} + w_z \frac{\partial t}{\partial z} = a_p \left(\frac{\partial^2 t}{\partial r^2} + \frac{\partial^2 t}{\partial z^2} + \frac{1}{r} \frac{\partial t}{\partial r} \right). \tag{4}$$

Inasmuch as the system of equations (1)–(4) involves four unknowns, it represents a closed system.

Let us formulate the uniqueness criteria. The following data enter into the uniqueness criteria:

1. The geometric characteristic of the region in which the process takes place. This region is determined by the surface of the melt, consisting of two parts: the surface of contact between the melt and walls of the crucible and the free surface of the melt and phase interface. If we assume a cylindrical crucible, the following relation applies to the first surface:

$$S_c = 2\pi r_0 h + \pi r^2_0, \tag{5}$$

where h is the height of the level of the melt, a time function, r_0 is the radius of the crucible.

The second surface is determined from the formula

$$S = \pi r_0^2$$

and is independent of time.

2. The initial conditions (since the thermal process may be assumed regular once it has spread over the entire diameter, the initial conditions drop out).

3. The boundary conditions, which specify the unknown functions at the boundary of the melt. Let us formulate the relations between the unknown functions at the surfaces of the melt and crucible. As a consequence of the melt particles adhering to the solid surface S_c of the crucible, the projections of the velocity of motion of the liquid phase at any point of this surface and at any instant are equal to zero, i.e., $w_r = w_z = 0$. We may assume also that $w_r = w_z = 0$ at the surface of the melt. The temperature of the liquid phase at any point of the surface S_c is equal to the temperature of the surface at this point, i.e., $t_m = t_w$, where t_w is the temperature at a fixed point of the surface S_c. We write the heat balance equation for the melt under steady-state conditions:

$$\int -\lambda_m \left| \frac{\partial t}{\partial n} \right|_m dS_{fr} + \int -\lambda_m \left| \frac{\partial t}{\partial n} \right|_{cr} dS_\kappa = q S_c =$$

$$= \int -\lambda_m \left| \frac{\partial t}{\partial n} \right|_w dS_w, \tag{6}$$

where q is the specific heat flux through the crucible wall, S_{fr} is the free surface of the melt, calculated from the relation

$$S_{fr} = \pi r_0^2 - \pi r_\kappa^2.$$

At the liquid–solid phase interface the unknown variables are related by the material balance equation

$$\int_0^{r_\kappa} 2\pi r_\kappa dr_\kappa w_\kappa \gamma_{cr} = \frac{dh}{d\tau} \pi r_0^2 \gamma_{p_0} \tag{7}$$

and the heat balance equation

$$-\int_0^{r_{\text{к}}} \lambda_p \frac{\partial t_p}{\partial z} 2\pi r_{\text{к}} dr_{\text{к}} + \omega_{\text{к}} q_{\text{cr}} \gamma_{\text{cr}} w_{\text{к}} = \int_0^{r_{\text{к}}} \lambda_{\text{к}} \frac{\partial t_{\text{к}}}{\partial z} 2\pi r_{\text{к}} dr_{\text{к}}, \tag{8}$$

where q_{cr} is the heat of crystallization, w_k is the crystal growth rate, r_k is the radius of the crystal, ω_k is the cross-sectional area of the crystal at the interface.

The Laplace equation is also valid at the interface:

$$p' - p = \sigma\left(\frac{1}{R_1} + \frac{1}{R_2}\right),$$

where σ is the surface tension, R_1 and R_2 are the principal radii of curvature at a given point on the surface.

It is known from differential geometry [2] that the sum $1/R_1 + 1/R_2$ represents twice the mean radius of curvature of the surface. Employing the expression for this quantity and transforming to cylindrical coordinates, we can write for the symmetry of the problem

$$p' - p = \sigma \frac{\frac{\partial^2 h}{\partial r_n^2} + \frac{1}{r} \frac{\partial h}{\partial r_n} \left[1 + \left(\frac{\partial h}{\partial r_n}\right)^2 \right]}{\left[1 + \left(\frac{\partial h}{\partial r_n}\right)^2 \right]^{3/2}}. \tag{9}$$

We obtain similarity invariants from the defining system of equations and uniqueness criteria. From Eq. (1)

$$\frac{z_0}{r_0}, \quad \frac{w_{z0}}{w_{r0}},$$

the equations of motion of the liquid phase (2) and (3) yield the following similarity invariants:

$$\frac{z_0}{r_0}, \quad \frac{w_{z0}}{w_{r0}}, \quad \text{Ho}_0, \quad \text{Fr}_0, \quad \text{Eu}_0, \quad \text{Re}_0,$$

where $\text{Ho}_0 = \omega_0 \tau_0/r_0$ is the homochronicity similarity criterion, $\text{Fr} = g r_0/w_0^2$ is the gravitational similarity invariant, $\text{Eu}_0 = p_0/\rho w_0^2$ is the force similarity invariant, $\text{Re}_0 = w_0 r_0/\nu_m$ is the dynamic similarity invariant. From Eq. (4) we obtain

$$\text{Ho}_0, \quad \text{Pe}_0, \quad \frac{z_0}{r_0}, \quad \frac{w_{z0}}{w_{r0}},$$

where $\text{Pe}_0 = w_0 r_0/a_{p_0}$ is the thermal similarity invariant.

From Eq. (5)

$$\frac{h}{r_0}, \quad \frac{s_c}{r_0^2},$$

from Eq. (6)

$$\frac{h}{r_0}, \quad \frac{r_{\text{к}}}{r_0}, \quad \frac{t_w}{t_0},$$

from Eq. (7)

$$\frac{r_{\text{к}}}{r_0}, \quad \frac{\gamma_{\text{к}\sigma}}{\gamma_{m_0}}, \quad \frac{w_{\text{к}}\tau_0}{h_0}, \quad \frac{h}{h_0}.$$

Equation (8) gives the following similarity invariants:

$$\frac{q_{cr}\gamma_{cr}w_{\kappa}r_0}{\lambda_{p_0}t_0}, \quad \frac{\lambda_{\kappa 0}}{\lambda_{p 0}},$$

Eq. (9) gives

$$\frac{\sigma}{p_0 r_0}.$$

On the basis of the π-theorem of similarity the dimensionless temperature may be written

$$\frac{t_m}{t_{m_0}} = \varphi\left(\frac{z}{h_0}, \frac{r}{r_0}, \frac{h}{r_0}, \frac{s_c}{r_0^2}, \frac{w_{z_0}}{w_{r_0}}, Ho_0, Fr_0, Eu_0, Re_0, Pe_0,\right.$$

$$\left.\frac{t_W}{t_0}, \frac{\sigma}{p_0 r_0}, \frac{w_{\kappa}\tau_0}{h_0}, \frac{\lambda_{\kappa 0}}{\lambda_{m_0}}, \frac{q_{cr}\gamma_{\kappa 0}w_{\kappa}r_0}{\lambda_{\kappa 0}t_0}, \frac{\tau}{\tau_0}\right). \tag{10}$$

In view of the fact that the velocities of motion of the melt are small, they may be disregarded. Furthermore, at the crystallization temperature it may be assumed that $\gamma_{k 0}/\gamma_{m 0}$ and $\lambda_{k 0}/\lambda_{m 0}$ are equal to unity. Then Eq. (10) assumes the form

$$\frac{t_m}{t_{m_0}} = \psi\left(\frac{z}{h_0}, \frac{r}{r_0}, \frac{s_c}{r_0^2}, \frac{t_W}{t_0}, \frac{w_{\kappa}\tau}{h_0}, \frac{q_{cr}\gamma_{\kappa 0}w_{\kappa}r_0}{\lambda_{\kappa 0}t_0}, \frac{\sigma}{p_0 r_0}\right). \tag{11}$$

During growth of the crystal a column of the melt is entrained behind it due to surface tension. Under definite thermal conditions crystallization occurs in the upper part of the liquid column. As a result, an object is obtained directly from the melt, its shape being determined by the shape of the upper part of the molten column. Consequently, control of the crystallization process reduces to control of the parameters of the molten column.

A number of problems relating to the study of capillary effects have been solved by Tsivinskii [3]. For the special case of a liquid column generated in the pulling of a circular rod from the free surface of a melt, he derived the expression

$$h_{co} = \sqrt{\frac{2\sigma}{\rho g}(1 - \cos\alpha_0) + \left(\frac{1}{2R_1^0}\right)^2 \frac{\sigma^2}{\rho^2 g^2}} - \frac{1}{2R^0} \cdot \frac{\sigma}{\rho g}, \tag{12}$$

where h_{co} is the height of the molten column, σ is the surface tension, ρ is the density of the melt, g is the gravitational acceleration, R^0 is the radius of curvature R_1 at $h = h_W$, α_0 is the slope of the tangent with respect to the vertical axis. We transform Eq.(12) so as to bring out the relations between the height h_{co} of the liquid column and the radius r_k of the pulled crystal. The radius R^0 is calculated from the formula used to calculate the radius of curvative R_1 for bodies of rotation:

$$R_1^0 = \frac{r_{\kappa}}{\sin\alpha_0}. \tag{13}$$

After the substitution of the resultant value of (13) into the expression (12), followed by a series of manipulations, we obtain the following expression relating h_{co} and r_k:

$$\frac{1}{2\sin^2\frac{\alpha_0}{2}} r_{\kappa}h_{co}^2 + \frac{\sigma}{\gamma}\cos^2\frac{\alpha_0}{2}h_{co} = \frac{\sigma}{\gamma}. \tag{14}$$

The ratio σ/γ will vary only slightly with temperature, so that it may be considered constant. Then, clearly, as the height of the column increases the radius r_k must decrease. With increasing temperature of the melt the height of the liquid column increases, implying a decrease in crystal diameter. An error is noted in Eq. (12), however. Tsivinskii supposes that the centers of curvature lie on the symmetry axis of the crystal. Actually, however, the centers of curvature are situated on an evolute that does not coincide with the crystal axis.

From the equation of material balance and the Laplace equation we can relate the derived complexes and simplexes at the phase interface by the following function on the basis of the π-theorem of similarity:

$$\frac{r_{\kappa}}{r_0} = \psi\left(\frac{\sigma}{p_0 r_0}, \frac{w_{\kappa}\tau}{h_0}, \frac{h_{CO}}{h_0}\right).$$

But since the height of the column h_{CO} is a function of the temperature, r_k/r_0 depends on the same simplexes and complexes as the temperature simplex t_m/t_{m0}, i.e.,

$$\frac{r_{\kappa}}{r_0} = \varphi\left(\frac{s_C}{r_0^2}, \frac{t_W}{t_0}, \frac{w_{\kappa}\tau}{h_0}, \frac{\sigma}{p_0 r_0}, \frac{q_{cr} \tau_{\kappa 0} w_{\kappa} r_0}{\lambda_{\kappa 0} t_0}\right). \tag{15}$$

The criterial relations (11) and (15) can be used to generalize the experimental data. As noted, experiments were performed on an apparatus with a standard shielding arrangement.

Ten holes were drilled in the upper shield to lead in thermocouples to the melt. The thermocouples were made of Chromel–Alumel wire 0.2 mm in diameter. The thermocouples, insulated by a quartz stem, were placed in quartz capillaries, which were rigidly secured in the upper shield through slotted graphite cones under pressure from graphite nuts (the nut was screwed down onto a bushing with the cone). This type of attachment was completely reliable and the position of the thermocouples was exactly fixed. Above the upper shield the thermocouples were inserted into a fiberglass stocking. The thermocouples were led out through an observation window in the furnace. The recording instruments were two twelve-point potentiometers, type ÉPP-09 from the Lenteplopriobor factory. Chromel–Alumel compensating wires were installed in the shield to eliminate possible noise in the potentiometer readings. The cold junctions of the thermocouples were immersed in ice water.

In each melting operation the temperature was measured at nine points of the melt at one height along the axis from the center of the crucible bottom. In repeated meltings the temperature values were obtained at three heights. Also measured in the experiments was the wall temperature of the rotating crucible. To accomplish this, thermocouples in quartz sheaths were sealed into the side wall and bottom of the crucible. The thermocouples were led out through the base of the crucible and a hollow steel shaft in the furnace. The thermocouple emf was transmitted to a potentiometer by means of a specially designed slip ring.

The flow of water coolant in all sections of the apparatus was held constant in all the tests.

Crystals of a specified diameter were obtained by manual regulation of the voltage.

During melting of the metal the crucible was situated in the lower position, so that the capillaries containing the thermocouples were located above the metal. After melting of the metal the crucible did not alter its position relative to the shields. The level of the melt during melting dropped 5 or 6 mm, however.

One of the objectives of the investigation was to establish the influence of the rate of ascent and rotation of the seed crystal on the temperature distribution in the melt for a given crystal diameter. Experiments were therefore performed under various pulling conditions:

n_{rot} = 20 rpm, 50 rpm, 93 rpm;

w_{asc} = 1.0 mm/min, 1.5 mm/min, 2.0 mm/min, 2.5 mm/min.

The crucible was rotated in all the tests at a constant rate n_c = 12 rpm.

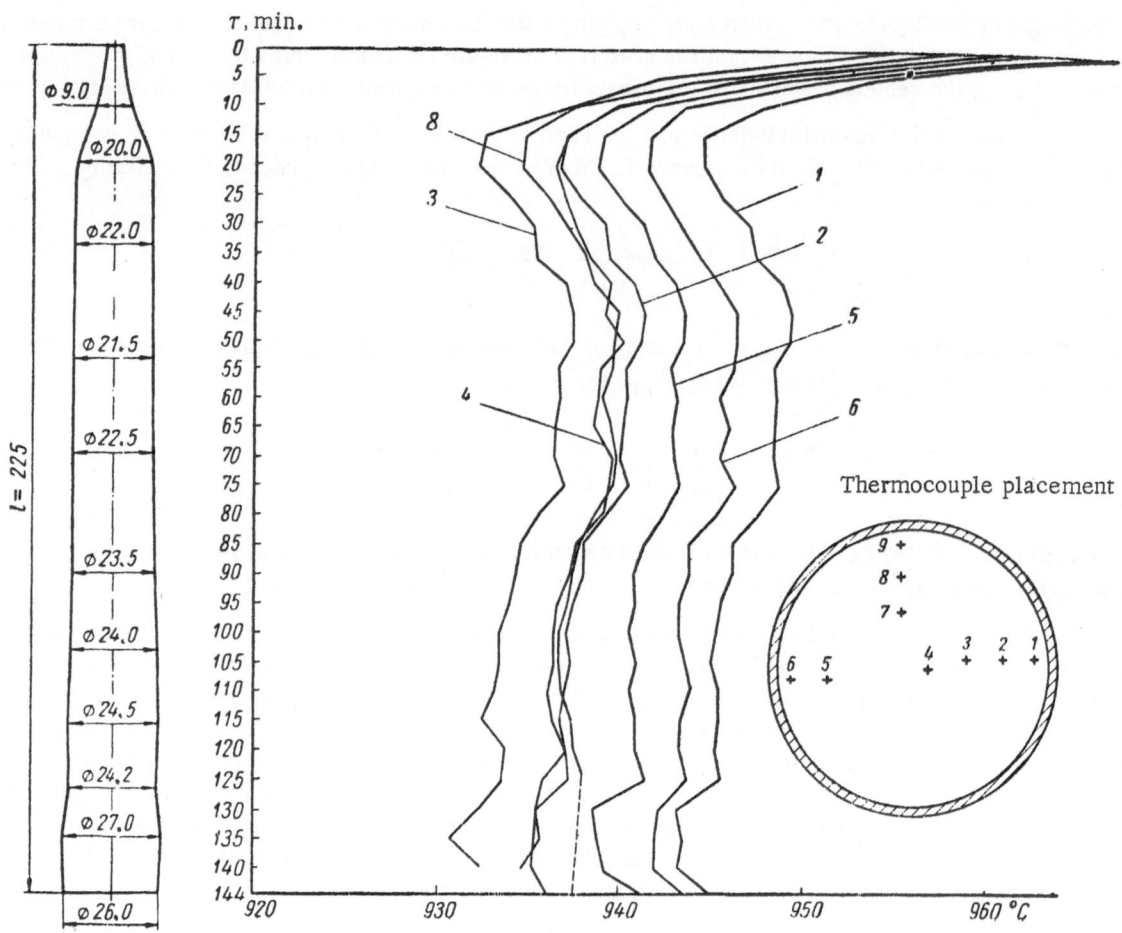

Fig. 1. Variation of temperatures in the melt during the growth process.

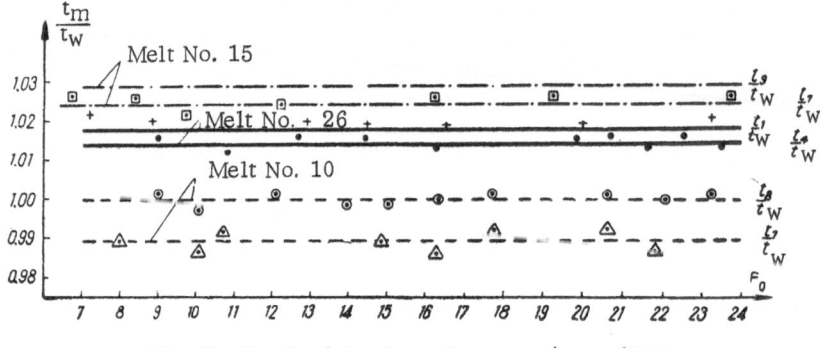

Fig. 2. Graph of the dependence $t_m/t_w = f(Fo)$.

During the experiments ingots were obtained with a mean diameter of 25 mm and length of 220 mm. The melting was carried out in an argon medium at a pressure $p_0 = 0.25$ atm. On one part of the ingots the electrical resistance was measured on the ends, sides parallel to the axis, and dislocations. On another part the form of the crystallization front was determined by anodic etching.

The experimental data were subjected to primary processing as follows. Graphs were constructed for the variation in temperature of the melt and crucible over the length of the crystal, showing the dependence of the ingot diameter on the true values of the melt temperatures. It is evident from the graph (Fig. 1) that as the temperatures of the melt increase the diameter of the ingot decreases. After processing in dimensionless (criterial) form (ingots with variable diameter from 20 to 30 mm), graphs were constructed for

$$\frac{t_m}{t_w} = f(\text{Fo}),$$

where Fo is the Fourier number, t_m is the temperature of the melt, t_w is the temperature of the crucible wall.

Under all conditions the dependence of the relative temperature of the melt t_m/t_w on the Fourier number turned out to be linear, with the straight line $t_m/t_w = \varphi(\text{Fo})$ parallel to the Fo axis (Fig. 2). Consequently, the relative temperatures of the melt do not depend on the Fourier number, thus demonstrating the self-similarity of the thermal field in the melt. Therefore, the relative temperature of the melt does not depend on the heater power (or on the ingot diameter).

With further processing the influence of rotation of the seed crystal on the temperature distribution in the melt becomes evident. As the rpm of the seed crystal is increased the relative temperature of the melt increases, this increase becoming more pronounced as the speed of rotation of the seed is varied from 20 to 50 rpm, less appreciably from 50 to 90 rpm.

Consequently, as a result of preliminary processing of the experimental data on the temperature field of the melt in one piece of equipment, the following conclusions may be drawn:

a. The temperature field of the melt is self-similar.

b. The diameter of the ingot depends on the values of the temperatures of the melt.

c. The temperature distribution in the melt is affected by the rate of rotation of the seed crystal.

Literature Cited

1. D. T. J. Hurle, Mechanism of Growth of Metal Single Crystals from the Melt [Russian translation] (IL, 1963).
2. V. I. Smirnov, Course in Higher Mathematics (OGIZ, 1948).
3. S. V. Tsivinskii, Application of the Theory of Capillary Effects to the Preparation of Specimens of Predetermined Form Directly from the Melt by the Method of A. V. Stepanov, Inzh.-Fiz. Zh., Nos. 5 and 9 (1962).
4. P. K. Konakov, Similarity Theory and Its Application in Heat Engineering (Gosénergoizdat, 1959).
5. C. Elbaum and B. Chalmers, Can. J. Phys., 33:196 (1955).